T0134424

Fuzzy Logic

Jenny Carter · Francisco Chiclana ·
Arjab Singh Khuman · Tianhua Chen
Editors

Fuzzy Logic

Recent Applications and Developments

 Springer

Editors
Jenny Carter 🄳
Department of Computer Science
University of Huddersfield
Huddersfield, UK

Arjab Singh Khuman 🄳
Faculty of Computing
Engineering and Media
De Montfort University
Leicester, UK

Francisco Chiclana 🄳
Faculty of Computing
Engineering and Media
De Montfort University
Leicester, UK

Tianhua Chen 🄳
Department of Computer Science
University of Huddersfield
Huddersfield, UK

ISBN 978-3-030-66476-3   ISBN 978-3-030-66474-9   (eBook)
https://doi.org/10.1007/978-3-030-66474-9

This Springer imprint is published by the registered company Springer Nature Switzerland AG
The registered company address is: Gewerbestrasse 11, 6330 Cham, Switzerland

Preface

Many problems within industry and commerce can be modelled mathematically and/or statistically. However, in practice, applications modelled in this way often perform poorly. Implemented assumptions can cause developed solutions to lack a certain level of robustness. Standard control applications, for example, often work poorly under certain conditions or they are not smooth in their movement. Physical measurements are, by their nature, imprecise. They are only as good as the instrument doing the measuring. A 2 Kg bag of sugar is never 'exactly' 2 Kg for example. Yet, traditional mathematically based control solutions use such measurements as being precise. Experts make decisions with imprecise data in an uncertain world. They work with knowledge that is rarely defined mathematically or algorithmically but uses vague terminology with words.

Fuzzy logic relies on the concept of a fuzzy set, which was proposed by Lotfi Zadeh, in his 1965 seminal paper—'Fuzzy Sets' (published in Information and Control, volume 8, pp. 338–353). Zadeh was a Professor at the University of Southern California until his death in 2017. The idea of fuzzy sets described in his seminal work lays the basis for Fuzzy Logic. Fuzzy Logic is particularly good at handling uncertainty, vagueness and imprecision. This is very useful where a problem can be described linguistically (using words) or, as with neural networks, where there is data and you are looking for relationships or patterns within that data. Fuzzy Logic uses imprecision to provide robust solutions to problems. Applications of fuzzy logic are varied and include robotics, washing machine control, nuclear reactors, information retrieval, train scheduling, system modelling, camera focus, stock tracking.

The chapters in this book provide further insight into the wide range of approaches to problem-solving using fuzzy logic and illustrate these approaches over a wide variety of application areas.

The Editors

Huddersfield, UK   Jenny Carter
Leicester, UK   Francisco Chiclana
Leicester, UK   Arjab Singh Khuman
Huddersfield, UK   Tianhua Chen

Contents

Fuzzy Logic, a Logician's Perspective

Patrick Fogarty

Abstract Fuzzy logic arises from an attempt to manage the inherent vagueness there is in the language we use when discussing our world—it is a formal treatment of vague predicates. This chapter will describe how this formal structure has come about, from origins in philosophical thought, through the development of non-standard logics. It will explore, from a logician's perspective, useful tools using fuzzy set theories, such as Basic Fuzzy Logic (BL) and T-Norm Fuzzy logics, deployed in computer systems today. It is intended to detail the techniques used to set up such theories and to review the relationship that Logic bears to them. In conclusion, it is proposed that further suggested theoretical investigations might yield useful practical results.

Keyword Aristotle · Axiomatization · BL · Eubulides · Fuzzy logic · Hájek · History · Logic · Logician's perspective · Set theories · Sorites · Suggested theoretical investigations · T-norms · Vagueness · Wang's paradox · Zadeh

1 Introduction

The subject of this chapter is the foundations of Fuzzy Set Theory and Fuzzy Logic. When applying techniques in computer science it is not necessary to know their historical development. Like driving a car, it is not necessary to understand the workings of the internal combustion engine. On the other hand, understanding engines and their history can enhance our driving experience and one gains a broader appreciation of the car as an object created by human endeavour. Similarly, when we stand back and view Fuzzy Logic from a historical perspective, by examining its foundations we gain an overview that increases our ability to see relationships, and allows us to explore options for future innovation. This chapter is intended to give the reader a path through the literature to help gain a historical perspective. This is not intended to be a comprehensive review, rather to inspire further reading; [1] a single source text covering the historical development of Fuzzy Set Theory and Fuzzy Logic, is a

P. Fogarty (✉)
Doora, Portmagee, Co. Kerry V23 RX94, Ireland

© Springer Nature Switzerland AG 2021
J. Carter et al. (eds.), *Fuzzy Logic*,
https://doi.org/10.1007/978-3-030-66474-9_1

good place to start. My conclusions propose some 'blue sky' ideas which are ultimately intended to pique interest and encourage further thought. As with so many other subjects, it all starts with the Greeks and Aristotle.

2   Ancient Greece

Aristotle in his Metaphysics book IV Page 1597 7: 23 [2] presents a formulation of the law of the excluded middle:

> But on the other hand, there cannot be an intermediary between contradictories, but of one subject we must either affirm or deny any one predicate.

Aristotle was aware that there are things that are indeterminate or at least problematic when trying to determine truth or falsity.

> What is, necessarily is, when it is; and what is not, necessarily is not, when it is not. But not everything that is, necessarily is; and not everything that is not necessarily is not...I mean, for example: it is necessary for there to be or not to be a sea-battle tomorrow; but it is not necessary for a sea-battle to take place tomorrow, nor for one not to take place—though it is necessary for one to take place or not to take place...Clearly, then it is not necessary of every affirmation and opposite negation one should be true and the other false. For what holds for things that are does not hold for things that are not but may possibly be or not be; with these it is as we have said. Barnes [2] De Interpretatione Page 30 9: 23

This is a quite brilliant analysis and shows how Aristotle understood contingency and recognised that there is a subtlety in the analysis of truth and falsity when considering indeterminate predicates. Unfortunately, Aristotle never pursued the issues much further. The development of Logics capable of addressing the question of pre-determination arising from Aristotle's analysis had to wait until Łukasiewicz in the twentieth century. The motivation for his multivalued logics was precisely to remove the dependence of logic on necessity:

> Even then I strove to construct non-Aristotelian logic, but in vain. Now I believe I have succeeded in this. My path was indicated to me by antinomies, which prove that there is a gap in Aristotle's logic. Filling that gap led me to a transformation of the traditional principles of logic. Examination of that issue was the subject-matter of my last lectures. I have proved that in addition to true and false propositions there are *possible* propositions, to which objective possibility corresponds as a third in addition to being and non-being. This gave rise to a system of three-valued logic, which I worked out in detail last summer. That system is as coherent and self-consistent as Aristotle's logic, and is much richer in laws and formulae. That new logic, by introducing the concept of objective possibility, destroys the former concept of science, based on necessity. Possible phenomena have no causes, although they themselves can be the beginning of a causal sequence. An act of a creative individual can be free and at the same time affect the course of the world. Simons [3] Sect. 5.2 'Indeterminism and the Third Value'

Aristotle presented us with a version of logical formalism, syllogistic or classical logic, which cannot handle vague concepts. The vague is dismissed from the logical framework and not addressed in Aristotle's logic. His contemporary Eubulides, in contrast, considered semantic paradoxes and found them interesting to explore [4]. Aristotle and Eubulides illustrate that, from the very beginnings of philosophical thought, thinkers have been aware of predicative vagueness and have tried in different ways to address it.

Eubulides specifically studied the type of semantic vagueness manifested in the "Sorites" paradox—a paradox conceived or at least popularised by him in the 4th century BCE. The name comes from the Greek σωρεία meaning heap. The Sorites paradox asks, "when does a collection of grains of sand become a heap?" Obviously, you cannot reasonably call one grain of sand a heap, nor two, nor three; perhaps a thousand grains? When is it exactly that a collection of grains of sand becomes a heap? The answer to the question is vague - it is not clear that there is a definitive answer.

More formally this paradox can be stated following Wang's paradox as in [5]:

$$if\ n \text{ is small } then\ n + 1 \text{ is small}$$
$$0 \text{ is small}$$
$$\therefore \text{ all numbers are small by mathematical induction}$$

In the case of Eubulides' Sorites paradox, the predicate heap is vague and in the case of Wang the predicate small is vague. This creates a problem when trying to correctly assign truth to a proposition. Aristotle's contention in proposing the law of the excluded middle is that a proposition must either be true or false and that there is no other possibility.

In classical logic, truth is a bivalent attribute of a proposition. A proposition is either true or false—there are no other options. This does not allow for vague concepts, which of itself is not a bad thing; taking this route allows the development of logic for well-defined predicates with no vagueness, but it inhibits discussion of propositions that contain vague predicates.

Aristotle had determined, in his discussion of the sea-battle, that for future events the law of the excluded middle did not apply [2] De Interpretatione Page 30 9: 30. It is not that Aristotle and classical logicians did not know of vagueness and the semantic paradoxes that arise, rather they could see no useful way to implement a logic to process propositions containing vague or indeterminate predicates. An analogy can be drawn between the emergence of non-Euclidean geometries and the arrival of non-classical logics. It was not until new axiomatisations for set theories and logics that vagueness could be exploited in a practical way. Non-Euclidean geometries began to be worked on in the 1820s by Bolyai and Lobachevsky [6] Chap. 3 Sects. 30 and 31. By 1882, Pasch [7] had published an axiomatization and demonstrated the power of axiomatics to produce deductive geometries. The applicability of axiomatics to other areas of mathematics was seen to be a useful tool by Hilbert [7] and it was Hilbert's vision that set the stage for developments in set theory in the twentieth century, and ultimately to the development of fuzzy set theory and fuzzy logic.

3   Twentieth Century

Hugely influential on twentieth-century thought was the Principia Mathematica of Whitehead and Russell [8]. In constructing a mathematical logic they had three aims: firstly to reduce primitives to the minimum, i.e. minimise undefined ideas and undemonstrated propositions; secondly to achieve precise expressions of mathematical propositions; and thirdly, to resolve the paradoxes of symbolic logic and the theory of aggregates. For a discussion of paradoxes see [9] for aggregates [10]. The work undertaken by Whitehead and Russell [8], and others such as Frege [11], provided formal tools for investigations in logic and set theory that were to lead to advances in both areas of research. It was not until the twentieth century that logicians had the motivation and the tools to begin exploring non-bivalent truth values for propositions with vague predicates.

Waismann, a member of the Vienna Circle [12] said "[W]e should be blind if we did not admit that the ideas "true" and "false" are often employed in ways running counter to orthodox logic. Thus, we say, … "This is not entirely true", …—phrases which strongly suggest that we regard the ideas "true" and "false" as capable of gradation …. the matter can be looked upon as showing the rudiments of a novel type of logic." [13].

In the 1920s, Łukasiewicz [3] was one of the first logicians to reject the principle of bivalence and present a logic with multivalued truth as opposed to the classical bivalent approach. Łukasiewicz worked on Aristotle's sea-battle question and introduced a three-valued logic to resolve the paradox. Later this was extended to a logic with m values L_m and infinite truth values L_o. Others such as Post and Gödel ([14] Sect. 5 History of Many valued Logic) worked on similar schemas, but it is of special historical interest that Łukasiewicz worked on the link to Aristotle's concerns about vague or indeterminate propositions. Łukasiewicz provided new interpretations of the formal logical work of ancient philosophers and transformed how historical logic was perceived.

Following on from the work of Łukasiewicz there were many who looked at both multivalued logics and set theories with non-binary multivalued characteristic functions determining set membership. Notably, Skolem ([15] Page 290 'Some Remarks on Axiomatized Set Theory') investigated axiomatisation of set theory and looked at the relationship between multivalued logics and set theory. Axiomatisation of many valued logics was researched by Louise Schmir Hay [16] in the late 1950s and early 1960s.

4   Modern Origins of Fuzzy Logic

The late nineteenth and early twentieth century witnessed a blossoming of research into the foundations of mathematics evidenced by the work of Lobachevsky, Bolyai and Riemann in Geometry; Frege, Whitehead and Russell and Hilbert in set theory.

These researches saw more and more areas of mathematics axiomatized and gave rise to a great interest in metamathematics, symbolic logic, paradoxes and set theory. It was the work undertaken in these areas and the desire to produce a graduate course in the foundations of mathematics that inspired Kleenes to write his Introduction to Metamathematics [17]. This rising interest in axiomatizing multivalued logics, Kleene's Introduction to Metamathematics [18], and a knowledge of Lattice theory [19], inspired Lofti Zadeh to look at vague concepts and postulate fuzzy sets. Dieter Klaua [20] worked simultaneously on a similar project but it is Zadeh who is credited with the discovery or invention of fuzzy set theory [21] and fuzzy logic [22].

Zadeh's work is seminal in that, from the outset, he looked at the applicability of mathematics to the "domains of pattern recognition, communication of information, and abstraction." Zadeh [21] Page 338. This represented a turning point, and the accessibility of his work was a major factor in its widespread application.

5   What Is a Fuzzy Set?

Zadeh defines a fuzzy set as one where membership is determined by a graded characteristic function that assigns to each object a grade of membership. This grading can be achieved by assigning each possible grade of membership a number between zero and one [21] Page 339. The purpose of such a definition is to enable one to discuss class membership for objects with predicated characteristics that are vague, such as the class of tall people or the class of numbers much greater than one. Zadeh's aim is to provide a framework which is more general than that of ordinary sets with wide application.

The framework is set up as follows, using Zadeh's notation:

Where X is a space of points (Universe of Discourse) with elements x, $X = \{x\}$

A *fuzzy set* A in X has a characteristic membership function $f_A(x)$

where $\{f_A(x)|x \rightarrow [0, 1], x \in A\}$

With these definitions one can define the usual set properties for fuzzy sets:

The Empty set

A fuzzy set A is empty if and only if $f_A(x) = 0$ for all $x \in X$

Equality of sets

Fuzzy sets A and B are equal if and only if $f_A(x) = f_B(x)$ for all $x \in X$

Complement of a set

The complement of a fuzzy set A, A', is given by

$$F_{A'}(x) = 1 - f_A(x)$$

Containment

A is a subset of *B* if and only if $f_A(x) \le f_B(x)$

Union membership characteristic function

C is the union of *B* and *A* or in symbols $C = B \cup A$ if and only if

$$f_c(x) = \text{Max}\,[f_A(x), f_B(x)], \ x \in X$$

Intersection membership characteristic function

C is the intersection of *B* and *A* or in symbols $C = B \cap A$ if and only if

$$f_c(x) = \text{Min}\,[f_A(x), f_B(x)], \ x \in X$$

6 What Is Fuzzy Logic?

While set theory deals with how objects belong to a set or aggregation, logic deals with the truth of propositions. There is a relationship between the two, and in a sense a set theory is a logic that deals with the truth of the proposition "*x* belongs to *A*," where *x* is a member of *X* the universe of discourse and *A* is a class in *X*. Thus, a multivalued logic can be constructed by rejecting the principle of bivalence and adopting a graded truth value. Multivalued logics were examined by Łukasiewicz who was well aware of the paradoxes around vagueness [23]. However, while Łukasiewicz was focussed on the theoretical aspects alone, Zadeh was interested in the application of fuzzy logic in processing vague language. He defines Fuzzy Logic in the following way:

> The term fuzzy logic is used in this paper to describe an imprecise logical system, FL, in which the truth-values are fuzzy subsets of the unit interval with linguistic labels such as true, false, not true, very true, quite true, not very true and not very false, etc. The truth-value set, $\mathcal{T}$, of FL is assumed to be generated by a context-free grammar, with a semantic rule providing a means of computing the meaning of each linguistic truth-value in $\mathcal{T}$ as a fuzzy subset of [0, 1]. Zadeh [22] Page 407.

The concept is to apply formal logic to real-world imprecise language by allowing truth values in FL to have the following properties:

- everything is, or is allowed to be, partial, i.e. a matter of degree,
- everything is, or is allowed to be, imprecise (approximate),
- everything is, or is allowed to be, granular (linguistic),
- everything is, or is allowed to be, perception based.

cf. https://wi-consortium.org/wicweb/pdf/Zadeh.pdf.

Bearing in mind the relationship set intersection has to logical conjunction, the basic schema having been established for fuzzy sets, various logics can be produced using different t-norms to define the logical conjunction $C = A \wedge B$ where now f maps to a truth value defined in the interval [0, 1].

Using the Łukasiewicz t-norm, Conjunction becomes

$$f_c(x) = \text{Max}\,[f_A(x) + (f_B(x) - 1), 0], \ x \in X$$

and we generate Łukasiewicz logic.

Using the Minimum t-norm for intersection as above, we get Gödel–Dummett Logic

$$f_c(x) = \text{Min}\,[f_A(x) + f_B(x)], \ x \in X$$

Using a product—product fuzzy logic

$$f_c(x) = f_A(x).f_B(x), \ x \in X$$

BL or Basic Fuzzy Logic is the logic of the class of all continuous t-norms.

7 Reception

The idea of fuzzy set theory was not well received to start with. There was a perception that fuzzy logic led somehow to fuzzy thinking and that there was no way to usefully apply fuzzy thinking. As applications of fuzzy logic to problems began to bear fruit, the reception of the theory has changed and now fuzzy set theory and fuzzy logic are accepted as legitimate areas of study especially in computer science.

Hájek, Godo and Esteva [24] propose the difference between probability and fuzzy logic can be described in the following terms:

In our opinion any serious discussion on the relation between fuzzy logic and probability must start by making clear the basic differences. Admitting some simplification, we consider that fuzzy logic is a logic of vague, imprecise notions and propositions, propositions that may be more or less true. Fuzzy logic is then a logic of partial degrees of truth. On the contrary, probability deals with crisp notions and propositions, propositions that are either true or false; the probability of a proposition is the degree of belief on the truth of that proposition. If we want to consider both as uncertainty degrees we have to stress that they represent very different sorts of uncertainty (Zimmermann calls them linguistic and stochastic uncertainty, respectively).

And further

Formally speaking, fuzzy logic behaves as a many-valued logic, whereas probability theory can be related to a kind of two valued modal logic…Thus, fuzzy logic is not a "poor man's probability theory, as some people claim."

8 Conclusions

The indeterminate still provokes controversy which is no bad thing. Controversy implies that as a subject it has not stagnated as a petrified forest of dogmas, nor lost all attraction for new projects and researchers. Fuzzy Set Theory and Fuzzy Logic deal with the indeterminate and are therefore living and vibrant areas for further research. There remains work to be done on the foundations of fuzzy logic:

- Is it possible to reduce Fuzzy Logic to some form of probabilistic manipulation? Despite Hajek's et al's claim to the contrary, would it be more efficient and perhaps more correct to use probability as the basis for computation where indeterminacy is involved?
  (Zadeh believed that probability theory should be based on fuzzy logic: https://kmh-lanl.hansonhub.com/uncertainty/meetings/zadeh03vgr.pdf)
- Can Fuzzy calculations be trusted *in extremis*? With Fuzzy implementations is it possible to apply proof theory? [25]
- How does fuzzy logic apply or interact with quantum logic? There is a suggestion that fuzzy logic can approximate or emulate quantum computing on traditional hardware [26]. Can fuzzy logic be implemented on quantum computing hardware?

The future looks fuzzy.

References

1. Belohlavek, R, Dauben, J. W., & Klir, G. K. (2017). *Fuzzy logic and mathematics: A historical perspective.* Oxford: Oxford University Press.
2. Barnes, J. (Ed.). (1984). *The complete works of Arsitotle.* s.l.: Princeton University Press.
3. Simons, P. (2020). Jan Łukasiewicz. In E. N. Zalta (Ed.), *The stanford encyclopedia of philosophy.* Summer 2020 ed. s.l:Metaphysics Research Lab, Stanford University.
4. Moline, J. (1969). Aristotle, Eubulides and the sorites. *Mind, 78,* 393–407.
5. Dummett, M. (1975). Wang's paradox. *Synthese, 30,* 301–324.
6. Wolfe, H. E. (2013). *Introduction to non-euclidean geometry.* s.l.: Dover Publications.
7. Kennedy, H. C. (1972). The origins of modern axiomatics: Pasch to Peano. *The American Mathematical Monthly, 79,* 133–136.
8. Whitehead, A. N., & Russell, B. (1962). *Principia mathmatica to *56* (2nd ed.). s.l.: The Syndics of the Cambridge University Press.
9. Cantini, A., & Bruni, R. (2017). Paradoxes and contemporary logic. In E. N. Zalta (Ed.), *The stanford encyclopedia of philosophy.* Fall 2017 ed. s.l.: Metaphysics Research Lab, Stanford University.
10. Burge, T. (1977). A theory of aggregates. *Noûs, 11,* 97–117.
11. Frege, G., & Jacquette, D. (2007). *The foundations of arithmetic: a logical-mathematical investigation into the concept of number 1884.* s.l.: Pearson Education.
12. Uebel, T. (2020). Vienna circle. In E. N. Zalta (Ed.), *The stanford encyclopedia of philosophy.* Summer 2020 ed. s.l.: Metaphysics Research Lab, Stanford University.
13. Waismann, F. (1968). Are there alternative logics? In R. Harré (Ed.), *How i see philosophy* (pp. 67–90). London: Palgrave Macmillan UK.
14. Gottwald, S. (2020). Many-valued logic. In E. N. Zalta (Ed.) *The stanford encyclopedia of philosophy.* Summer 2020 ed. s.l.: Metaphysics Research Lab, Stanford University.

15. van Heijenoort, J. (2002). *From Frege to Gödel: A source book in mathematical logic, 1879–1931 (Source Books in the History of the Sciences).* s.l.: Harvard University Press.
16. Hay, L. S. (1959). *An axiomatization of the infinitely many-valued predicate calculus.* s.l.: Cornell Univ.
17. Kleene, S. C. (1991). The writing of introduction to metamathematics. In T. Drucker (Ed.), *Perspectives on the history of mathematical logic* (pp. 161–168). Birkhäuser Boston: Boston(MA).
18. Kleene, S. C., & Beeson, M. J. (2009). *Introduction to metamathematics.* s.l.: Ishi Press International.
19. Vaughan, H. E. (1940). Garrett Birkhoff. *Lattice theory* (vol. 25, 155 pp). New York: American Mathematical Society Colloquium publications, American Mathematical Society; *Journal of Symbolic Logic, 5,* 155–157.
20. Klaua, D. (1967). Ein Ansatz zur mehrwertigen Mengenlehre. *Mathematische Nachrichten, 33,* 273–296.
21. Zadeh, L. A. (1965). Fuzzy sets. *Information and Control, 1*(8), 338–353.
22. Zadeh, L. A. (1975). Fuzzy logic and approximate reasoning (In Memory of Grigore Moisil). *Synthese, 30,* 407–428.
23. Kotarbiński, T. (1958). Jan Łukasiewicz's works on the history of logic. *Studia Logica: An International Journal for Symbolic Logic, 8,* 57–63.
24. Montagna, F. (2014). Hájek Petr, Godo Lluis, and Esteva Francesc. A complete many-valued logic with product-conjunction. *Archive for Mathematical Logic, 35*(1996), 191–208; *Bulletin of Symbolic Logic, 9*(6), 346–347.
25. Metcalfe, G., Olivetti, N., & Gabbay, D. (2009). *Proof theory for fuzzy logics.* s.l.:s.n.
26. Hannachi, S., Hatakeyama, Y., & Hirota, K. (2007). Emulating qubits with fuzzy logic. *JACIII 1, 11,* 242–249.

A Fuzzy Approach to Sentiment Analysis at the Sentence Level

Orestes Appel, Francisco Chiclana, Jennifer Carter, and Hamido Fujita

Abstract The objective of this chapter is to present a hybrid approach to the Sentiment Analysis problem focused on sentences or snippets. This new method is centred around a sentiment lexicon enhanced with the assistance of SentiWordNet and fuzzy sets to estimate the semantic orientation polarity and intensity for sentences. This provides a foundation for computing with sentiments. The proposed hybrid method is applied to three different datasets and the results achieved are compared to those obtained using Naïve Bayes (NB) and Maximum Entropy (ME) techniques. It is demonstrated through experimentation that this hybrid approach is more accurate and precise than both NB and ME techniques. Furthermore, it is shown that when applied to datasets containing snippets, the proposed method performs similar to state-of-the-art techniques.

Keywords Sentiment analysis · Hybrid method · Fuzzy sets · Fuzzy methods · Machine learning · Computing with sentiments

O. Appel (✉) · F. Chiclana
Institute of Artificial Intelligence (IAI), De Montfort University, Leicester, UK
e-mail: orestes.appel@email.dmu.ac.uk

F. Chiclana
e-mail: chiclana@dmu.ac.uk

J. Carter
University of Huddersfield, Huddersfield, UK
e-mail: j.carter@hud.ac.uk

H. Fujita
Intelligent Software Systems Laboratory, Iwate Prefectural University, Takizawa, Iwate, Japan
e-mail: issam@iwate-pu.ac.jp

© Springer Nature Switzerland AG 2021
J. Carter et al. (eds.), *Fuzzy Logic*,
https://doi.org/10.1007/978-3-030-66474-9_2

11

1 A Fuzzy Approach to Sentiment Analysis

In this document, we will introduce our hybrid approach to the Sentiment Analysis (SA) problem. Our focus will be SA at the sentence level. This approach is based on a sentiment lexicon and fuzzy sets, with the objective of computing the semantic orientation of sentences, including not only their polarity but their intensity too. This approach provides a foundation for computing with sentiments. Our method is executed using three different datasets and the results achieved are compared to the outputs of two machine learning (ML) techniques, namely, Naïve Bayes and Maximum Entropy. In this report, we prove that our proposed method obtains a higher level of accuracy and precision than the ML techniques just mentioned when the object of the SA process is to analyse *snippets*.

2 Introduction

The key objective of SA is to find out the attitude that people have with regard to some issue or subject, and usually, that attitude belongs in one of the three following categories: *positive*, *negative* or *neutral/objective*.

In the last few years, machine learning has been the technique-of-choice to address SA challenges. More specifically, supervised machine learning. Considering *fuzzy sets* mathematical properties and ability to manage vagueness and uncertainty, we do believe that fuzzy sets are a great choice to model sentiment. Hence, a combination of techniques could be successful at addressing the SA challenges by exploiting the best that each method has to offer. In the next paragraph, we will address our motivation for exploring this.

In Dzogang et al. [8], we see that very often authors deal mainly with psychological models when addressing the SA challenge. Nevertheless, other models may as well be successful. "It must be underlined that some appraisal-based approaches make use of graduality through fuzzy inference and fuzzy aggregation for processing affective mechanisms ambiguity and imprecision." said Dzogang et al. Liu [18], one of the main world experts in the SA domain, says that "we probably relied too much on Machine Learning". When it comes to advances in SA techniques, Poria et al. [25] presented the idea of combining machine learning, common-sense computing and linguistics with the purpose of improving precision in polarity detection. This concept of bringing together different techniques into one is what we have called a *hybrid style*. As such, the following themes could be applied in conjunction:

– Expressing sentiment *graduality* by utilising *fuzzy sets*.
– The application of semantic rules and a sound *sentiment lexicon* to compute sentiment's polarity.

Before us, other researchers have explored the potential of hybrid approaches. Poria et al. in [25] explore a new framework for concept-level sentiment analysis which they called Sentic Pattern. The authors claim that "by allowing sentiments to flow from concept to concept based on the dependency relation of the input sentence, authors achieve a better understanding of the contextual role of each concept within the sentence and, hence, obtain a polarity detection engine that outperforms state-of-the-art statistical methods". When they cannot find a matching *sentic pattern* [5] they resolve to Supervised Machine Learning. The *hybrid approach* that we present in this article attempts to address the case when a given word is not found in its lexicon by using a dictionary of words' frequencies and occurrences in place of recurring to machine learning as in [5]. Our method becomes a foundational element of *computing with sentiments*. The latter is derived from Zadeh's original idea: computing with words [38]. For more information on Sentiment Analysis (SA), refer to the work of Appel et al. [2], and for a full review of the topic access the work of Ravi and Ravi [27].

3 Research Methodology

In the devising of our hybrid approach, we established that its performance should match or exceed the level achieved by accepted supervised machine learning classification solutions. As such, the results of our method are compared against those achieved using Naïve Bayes (NB) and Maximum Entropy (ME). As we are focusing our system on SA at the sentence level, the latter is the right decision because as stated by Wang and Manning [32], Naïve Bayes actually outperforms Support Vector Machine for 'snippets': "[...] for short snippet sentiment tasks, NB actually does better than SVMs (while for longer documents the opposite result holds).". The comparison will be centred around *sentiment polarity determination*.

As part of our research, we went as well through the process of identifying the proper datasets to use in the comparison process. Those datasets are

- A Twitter corpus (Sentiment140), available at http://help.sentiment140.com/for-students. For identification purposes this dataset will be labelled as **Twitter A**.
- A Twitter Sentiment Analysis Training Corpus, available at http://thinknook.com/twitter-sentiment-analysis-training-corpus-dataset-2012-09-22/ that includes around 1.5 million tweets. In turn, this dataset is inspired by an SA competition promoted in Kaggle by the University of Michigan and a Twitter Sentiment Corpus generated by Niek Sanders. For identification purposes, this dataset will be labelled as **Twitter B**. *Note*: from this dataset, we have chosen randomly 1,000 tweets of each type (positive polarity and negative polarity).

In addition, we will use the Movie Review Dataset provided by Pang and Lee that is available at http://www.cs.cornell.edu/people/pabo/movie-review-data/. Many researchers have used this dataset [22] in their work, which makes it a very good choice for benchmarking purposes.

In terms of indicators to be utilised for measuring performance, we will adhere to widely used indices [12, 28]:

- *Accuracy*: the portion of all true predicted instances against all predicted instances.
- *Precision*: the portion of true positive predicted instances against all positive predicted instances.
- *Recall*: the portion of true positive predicted instances against all actual positive instances.
- *F1-score*: a harmonic average of precision and recall.

Readers are pointed towards the work of Sadegh et al. [12, 28] for a thorough definition of these aforementioned metrics.

4   A Hybrid Approach to the SA Problem at the Sentence Level

As mentioned above, the method to be presented is a 'hybrid approach' from two different perspectives. Namely, (a) the techniques utilised by the opinion/sentiment classifier and (b) the algorithms applied to articulate core components in the proposed method, as the generation and population of the sentiment lexicon and the dictionaries containing word-frequency occurrences. See Fig. 1 for the schematics of our proposed solution.

4.1   Component 1: The Sentiment/Opinion Lexicon

Based on the opinion lexicon compiled by Liu [15], which contains a list of 6,800 sentiment-carrying words, we started building our sentiment lexicon. This base lexicon needed enrichment by adding *polarity or valence values/scores*. For that purpose, we matched the terms in our lexicon with those existing in *SentiWordNet*, the latter created by Esuli and Sebastiani [9–11].

It must be clarified that the words in the lexicon are limited to part-of-speech items capable of conveying sentiments. Those are nouns, verbs, adjectives and adverbs, as established by a number of linguists and other experts [13, 14, 17, 33]. The polarity scores extracted from SentiWordNet contain values in the interval [0, 1]. As such, we can establish the following:

$$0 \leq PositiveValue, \; NegativeValue, \; ObjectivityValue \leq 1$$

$$0 \leq (PositiveValue + NegativeValue) \leq 1$$

$$ObjectivityValue = 1 - (PositiveValue + NegativeValue)$$

Hence, when the addition of *PositiveValue* (PSC) and *NegativeValue* (NSC) is equal to 1 for a given term $Word_k$, then such a word, $Word_k$, is fully opinionated, as

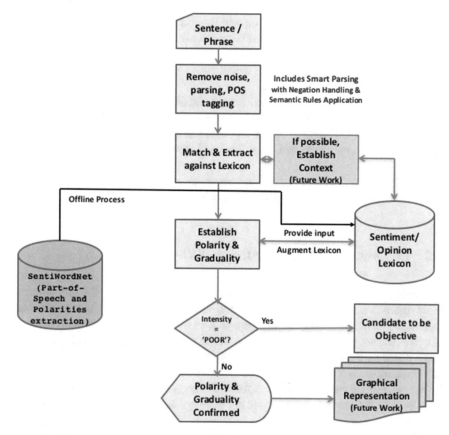

Fig. 1 Schematics of the proposed hybrid approach

opposed to when the sum of these two scores is 0. In such a situation the word, $Word_k$, is considered to be of a Neutral/Objective nature. In essence, the *ObjectivityValue* (COBJ) can be understood as a representation of ambiguity or hesitancy, as it is computed by obtaining the difference between 1 and the classification of a word as a negative/positive carrier of meaning. On the other hand, if PSC and NSC add, for the sake of the argument, to 0.8, then there are 0.2 points for the given word to 'semantically represent a neutral space of hesitancy'.

There were also hard challenges, as *not* every word in Liu's opinion lexicon was included in SentiWordNet. However, for those absent words in the lexicon we decided to keep them. Nevertheless, as they did not have neither polarity scores nor part-of-speech tags, we did mark them so they could be clearly identified and enhanced later on, as soon as the 'required' data became available. A full description of the elements of the Sentiment/Opinion Lexicon follows:

Word: an entry (word) in our sentiment lexicon.

SOL: semantic orientation (positive/negative); used "as is" from Hu and Liu lexicon [15].

PoS: part-of-speech, where each particle $\in$ {*noun, verb, adjective, adverb, satellite adjective*}

PSC: Positive Value/Score as extracted from SentiWordNet [10].

NSC: Negative Value/Score as extracted from SentiWordNet [10].

COBJ: Computed Neutral/Objectivity Value or Score.

VDX: Lexicon term version, used for the identification of synonyms (for future use).

UPDC: Update Counter: utilised to keep score of every opportunity in which a word in the lexicon gets updated with enhancements.

4.2 Component 2: Semantic Rules (SR)

Semantic rules are essential as they heavily assist in modelling the SA problem in a more rigorous way. Several authors like those authoring [21, 31, 34] have established that negation rules handling and the management of part-of-speech particles of interest, like 'unless, despite, but, ...' are key. These aforementioned research efforts are exemplified by the work of Xie et al. in [34], which includes a complete description of a semantic rules-based trend. The semantic rules devised and implemented in our hybrid system are partially based on those designed by Xie et al. A subset of these rules borrowed from [34] were used in our solution, and then they were *enhanced* by new rules that our team added. We have applied almost the same rules naming style used by Xie et al. but we added a sub-index (Rk_{HSC}) to represent those rules in [34] that we actually employed in our hybrid method. There are gaps in the numbering discussed by Xie et al. and they correspond to rules that were not implemented in our solution, for example (R9, R8, R5, R4 and R2). Tables 1 and 2 present the semantic rules used in our solution.

Notice the enhancement we have provided by adding *two new rules* for dealing with particular part-of-speech particles that were not incorporated in the original rules presented in [34]. Those are the word **while** and the word **however**. The newer rules are presented in Table 3.

It is well known in the linguists' community that words under the scope of negation may behave in unexpected ways. As per Potts [26], the 'Weak' words—such as *good* and *bad* behave like their opposites when negated—while 'Strong' words like *superb* or *terrible* have very generic semantics under negation. As per Potts [26], "not superb is consistent with everything from *horrible* to *just-shy-of-superb*, and different lexical items for different senses. These observations suggest that it would be difficult to have a general a priori rule for how to handle negation". It does not just turn good to bad and bad to good, Potts continues [26]: "Its effects depend on the words being negated. An additional challenge for negation is that its expression is lexically diverse and its influences are far-reaching (syntactically speaking)". Potts suggests methods

Table 1 Semantic rules actually implemented in our Hybrid Approach (HSC)

Rule	Semantic rules	Example
$R1_{HSC}$	"Polarity (not var_k) = -Polarity (var_k)"	"not *bad*"
$R3_{HSC}$	"Polarity ($NP_1\ VP_1$) = Compose (NP_1, VP_1)"	"*Crime* has decreased"
$R6_{HSC}$	"Polarity (ADJ to VP_1) = Compose (ADJ, VP_1)"	"*Unlikely* to destroy the planet"
$R7_{HSC}$	"Polarity ($VP_1\ NP_1$) = Compose (VP_1, NP_1)"	"*Destroyed* terrorism"
$R10_{HSC}$	"Polarity (not as ADJ as NP) = -Polarity (ADJ)"	"That wasn't as *bad* as the original"
$R11_{HSC}$	"If sentence contains 'but', disregard previous sentiment and only take the sentiment of the part after 'but'"	"And I've never liked that director, *but* I loved this movie"
$R12_{HSC}$	"If sentence contains 'despite', only take the sentiment of the part before 'despite'"	"I love the movie, *despite* the fact that I hate that director"
$R13_{HSC}$	"If sentence contains 'unless', and 'unless' is followed by a negative clause, disregard the 'unless' clause"	"Everyone likes the video *unless* he is a sociopath"

Table 2 Compose function implemented in HSC

Compose functions revised	Algorithms
"Compose (arg1, arg2)"	1. "Return -Polarity(*arg2*) if *arg1* is negation"
	2. "Return Polarity(*arg1*) if (Polarity(*arg1*) = Polarity(*arg2*)"
	3. "Otherwise, return the majority term polarity in *arg1* and *arg2*"

Table 3 New semantic rules extending those presented by Xie et al. in [34]

Rule	Semantic rules	Example
$R14_{HSC}$ (addition)	"If sentence contains 'while', disregard the sentence following the 'while' and take the sentiment only of the sentence that follows the one after the 'while'"	"'*While*' they did their best, the team played a horrible game"
$R15_{HSC}$ (addition)	"If sentence contains 'however', disregard the sentence preceding the 'however' and take the sentiment only of the sentence that follows the 'however'"	"The film counted with good actors. '*However*', the plot was very poor"

introduced by Das and Chen [7] and Pang et al. [23] for approximating the effects of negation; when the aforementioned technique is used at the tokenization level, it does translate into well-managed negation. In order to deal with negation and its long-distance effects, we have implemented this technique in our *proposed method*, and we have called it *smart tokenization*.

4.3   Component 3: Fuzzy Sets Approach to the SA Problem

In this sub-section, we present the strategies needed to classify sentences into either *Positive* or *Negative*. Furthermore, we qualify the *strength* of this previously identified polarity. For that, we proceed as follows:

– Define and produce the *fuzzy methodology* that will be applied
– Provide details of the fuzzy granulation—or the linguistic discrimination—that will be used to represent the subjective classification of sentences into either positive or negative
– Create the logic necessary that will jointly operate with the sentiment lexicon and the fuzzy sets presented in the top item
– Describe the mechanics of the whole classification process as we add the application of fuzzy sets.

4.3.1   Definition of Perceptions and Linguistic Variables for Polarity Intensity

As per Lotfi Zadeh, humans continually use *perceptions*. Quoting Zadeh [37]: "reflecting the bounded ability of the human brain to resolve detail, perceptions are intrinsically imprecise. In more concrete terms, perceptions are f-granular, meaning that (1) the boundaries of perceived classes are unsharp and (2) the values of attributes are granulated, with a granule being a clump of values (points, objects) drawn together by indistinguishability, similarity, proximity and function" (see Fig. 2 for a re-illustration of the graphic originally published by Zadeh in [38]). In another Zadeh's writing [38], he continues, by saying that "a granule may be crisp or fuzzy, depending on whether its boundaries are or are not sharply defined. For example, age may be granulated crisply into years and granulated fuzzily into fuzzy intervals labeled very young, middle-aged, old and very old." Figure 3 re-illustrates the graphical representation of the latter idea as originally presented by Zadeh in [37]. In 1973, Zadeh presented his thoughts on what he called linguistic variables: "a variable whose values are words instead of numbers" [36]. When it comes to which *linguistics variables* to use in our sentiment analysis problem, we realised that the *intensity* or *strength* with which the 'positivity' or 'negativity' of a sentence S could be understood corresponds to a *perception* as described by Zadeh. The perception P_S that a given person Y has about how positive or negative a sentence S might be.

Fig. 2 The concept of a granule as presented by Zadeh

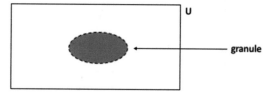

Informal: a granule is a clump of objects (points) drawn together by indistinguishability, similarity, proximity or functionality

U

granule

Formal: a granule is a clump of objects (points) defined by a generalized constraint

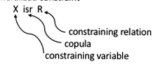

X isr R

constraining relation
copula
constraining variable

Fig. 3 Crisp granulation and fuzzy granulation as introduced by Zadeh

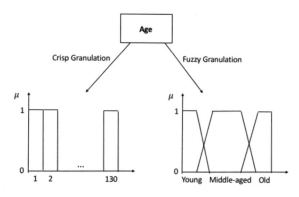

A sentence could either be *Negative* or *Positive*, and then, on top of it, 'Most Positive' or 'Very Positive', or 'Most Negative' or 'Very Negative', or something similar. Based on the definitions and concepts provided by Zadeh, a fuzzy granulation of positive/negative sentiments using fuzzy intervals is considered to be appropriate to deal with the problem at hand. According to Miller [20], *7 plus or minus 2*, corresponds to the effective number of categories that a given individual subject can handle. We selected and devised *5 labels* that correspond to $(7 - 2)$, distributed symmetrically in the domain $[0, 1]$. Notice that the choice of using a *trapezoidal function* is related to the ability of the latter to generalise triangular functions and we have aimed for more generality and for having more than one value *only* at the top of each category. A trapezoidal membership function, as displayed in Fig. 4, is represented by the 4-tuple (a, b, c, d).

Furthermore, the granules on the perception of the *negativity* or *positivity* of a sentence X are presented as: $G = \{Poor; \ Slight; \ Moderate; \ Very; \ Most\}$, by the following 4-tuple group:

- $(0, 0, 0.050, 0.150)$ Membership Function is instantiated to Granule Label = *Poor*

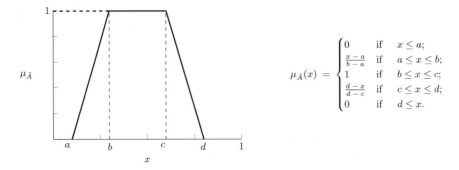

$$\mu_{\tilde{A}}(x) = \begin{cases} 0 & \text{if} & x \leq a; \\ \frac{x-a}{b-a} & \text{if} & a \leq x \leq b; \\ 1 & \text{if} & b \leq x \leq c; \\ \frac{d-x}{d-c} & \text{if} & c \leq x \leq d; \\ 0 & \text{if} & d \leq x. \end{cases}$$

Fig. 4 Trapezoidal membership function

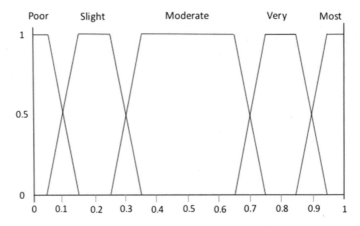

Fig. 5 Linguistic variables, fuzzy granulation and trapezoidal membership functions

– (0.050, 0.150, 0.250, 0.350) Membership Function is instantiated to Granule Label = *Slight*
– (0.250, 0.350, 0.650, 0.750) Membership Function is instantiated to Granule Label = *Moderate*
– (0.650, 0.750, 0.850, 0.950) Membership Function is instantiated to Granule Label = *Very*
– (0.850, 0.950, 1, 1) Membership Function is instantiated to Granule Label = *Most*

This way, the *intensity* related to the semantic positive or negative value for any word is assigned a certain fuzzy interval as illustrated in Fig. 5. From now on, we can compute the level of intensity related to the polarity of a sentence, or how strong or weak a given positive/negative sentiment might be as expressed in English. Furthermore, we are capable of saying that the sentiment of an individual towards a given sentence is, for example, *moderately* positive/negative, *poorly* positive/negative or *most* positive/negative, as the linguistic labels we have already introduced above (Fig. 6).

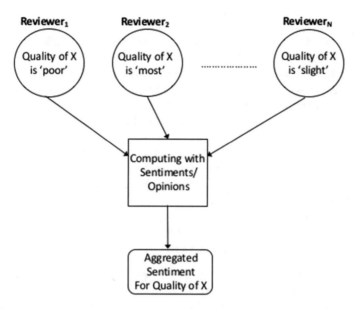

Fig. 6 Generic diagram of *computing with sentiments*, inspired by Zadeh

4.4 Description of the Process Implemented in Our Hybrid Approach

In this part of the document, we will describe the mechanics of our proposed classification system that will compute both, the sentiment polarity of a given sentence S and its intensity. The process will be executed in two steps which will be addressed below.

4.4.1 Hybrid Standard Classification (HSC): The Computation of the Sentiment's Polarity

The tasks described here are order-sensitive. The output of step 1 is used as input to step 2, the output of step 2 becomes the input of *step 3* and so on until the list of steps to execute is exhausted.

1. Step 1: tokenization, past-of-speech tagging, error handling and smart parsing. All these are achieved by the orderly application of the semantic rules presented in Sect. 4.2. When applicable, composed sentences are identified so that the adequate tagging is executed, so that the complete polarity is calculated as per the appropriate composition rule (Table 2), once *interpretation time* arrives. The smart tokenization process to handle negation that we described above is applied at this point in time, too.

2. Step 2: conveying sentiment terms, like adjective, nouns, verbs and adverbs, that are identified are then looked-up in the sentiment lexicon, fetching the matching parts-of-speech and polarity values.
3. Step 3: terms that are found *absent* in the Sentiment Lexicon are properly tagged. If needed, those words will be *enriched* in a separate sub-process aimed to expand the capabilities of the sentiment lexicon.
4. Step 4: the computation of the semantic orientation (SOR) of each sentence in the dataset is performed.
5. Step 5: once step 4 is completed, we treat as exceptions those terms for which a semantic orientation label (positive/negative) have not been found (either because the term was not included in SentiWordNet or it was there but it lacked polarity values.
6. Step 6: the final semantic orientation is generated for compounded sentences, by grouping together its compounded semantic orientation (CSO) based on its sub-sentence semantic orientation values, or $SORs$.

Let us demonstrate the above with an example. When computing the semantic orientation of a sentence (SOR), the positive/negative tag associated with the words in the sentence at hand that happen to be in the sentiment lexicon, as well as its polarity scores, are applied. Our proposed system executes a word counting process of semantic orientations for every sentence.

If *count*(positive words) $>$ *count*(negative words)
then [the sentence is classified as 'positive'], hence SOR = 'Positive'

If *count*(positive words) $<$ *count*(negative words)
then [the sentence is classified as 'negative'], hence SOR = 'Negative'

If *count*(positive words) $=$ *count*(negative words)
then [There is a tie. Follow alternative process], hence SOR = Table 4 result.

This stratified algorithm which consists of three levels, as shown in Table 4, resolves existing ties. As the strata displayed are mutually exclusive against each other, then a given stratus is executed *if and only if* the previous stratus is incapable of resolving an existing tie. Keep in mind that the intensity polarity (IP) scores in our sentiment lexicon range in the interval $[0, 1]$, and that the semantic orientation computation we have implemented requires both, the Positive/Negative tags and the positive/negative IPs in our sentiment lexicon. If a sentence S is made of n sub-sentences $(S_1, S_2, \ldots, S_n)$, then the CSO of the full paragraph/sentence is calculated by SOR sub-sentence counting.

1. **If** *count*(Positive SOR Sentences) $>$ *count*(Negative SOR Sentences)
   then $CSO_{(S_1, S_2, \ldots, S_n)}$ = Positive
2. **If** *count*(Positive SOR Sentences) $<$ *count*(Negative SOR Sentences)
   then $CSO_{(S_1, S_2, \ldots, S_n)}$ = Negative

Table 4 Tie-breaking process—a stratified algorithm

Stratus No.	Description
1	Review of polarity and intensity values/scores (IP). The highest value wins
2	If the previous step is unsuccessful in producing a classification output, then a hierarchy of importance among the several part-of-speech particles of interest is established. That hierarchy ranges from *most influential* to *least*: adjectives first, then adverbs followed by verbs and, finally, nouns. When needed, this hierarchy just steps in and a higher priority is assigned accordingly
3	If after applying two steps above there is still no resolution, we proceed by looking into the word-dictionary and searching for all participant words. Then, we extract the frequencies with which each word has appeared with a specific polarity, either positive or negative, and the polarity associated to the highest value wins

3. **If** $count$ (Positive SOR Sentences) $= count$ (Negative SOR Sentences)
   then $CSO_{(S_1, S_2, \ldots, S_n)} = SOR$ of S_k; $IP(S_k) = \max\{IP(S_1), IP(S_2), \ldots, IP(S_n)\}$

Keep in mind that the separators for sentences are regular English punctuation marks (PM):

$PM_i \in$ {period, comma, exclamation sign, question mark, colon, semicolon, ...}, where $i = 1 \ldots n$.

The sentences would then be broken into sub-sentences at tagging/parsing time. For long paragraphs/sentences, there would be many sub-sentences participating in many compositions. For short paragraphs—better known as snippets—like those available in the Twitter datasets we have utilised, we would have to compute *compose semantic orientations* for a lower number of sub-sentences. By looking at samples of the data in the Twitter database, we notice that more often than not there are no punctuation marks being used, which resulted in no generation of sub-sentences at tokenization time.

4.4.2 Hybrid Advanced Classification (HAC): The Computation of the Intensity of Polarity

This second step of our algorithm adds to the standard classification process by augmenting its capabilities by

1. Computing the polarity intensity (IP) with which a sentence X leans towards being *positive* or *negative*.
   The polarity intensity IP of a sentence (X) is to be obtained from the IP values of its associated list of sentiment-capable words ($W_1, \ldots, W_n$). Hence, the partial IP values of words of a sentence X are to be fused completely

24

O. Appel et al.

to produce the global sentence IP score. Formally, this means that an proper mapping $f : [0, 1]^n \rightarrow [0, 1]$ is required to be defined in a way that

$$IP(X) = f\big(IP(W_1), ..., IP(W_n)\big).$$

These so-called *fusion operators* can be classified into the following types: *conjunctive*, *disjunctive* and *compensative*. Their descriptions, shown below, have been taken from Appel et al. [3]:

(a) "Conjunctive operators behave like a logical 'and'. In this case, the global IP scores would be high only when all the partial IP values are high, too, but compensation is not possible as the presence of just one small partial IP value will result in a small global value no matter how big the rest of partial IP values are. A well-known family of conjunctive operators in the fuzzy realm is the t-norm family, and the minimum operator is the largest of all t-norms".

(b) "Disjunctive operators behave like a logical 'or", and can be seen as the dual operators of conjunctive operators. In this case, the global IP is low only when all the partial IP values are low. As with conjunctive operators, compensation is not possible as the presence of just one high partial IP value will result in a high global IP score, no matter how small the rest of partial scores are. In the fuzzy world, the family of t-conorms belongs to this type of operator, and the maximum is the smallest of all t-conorms".

(c) "Compensative operators are placed between the minimum and the maximum, and therefore they are neither conjunctive nor disjunctive. In this kind of operator, a small partial IP value can be compensated by a high partial IP value. This type of operator is also known as an *averaging operator*, with mean, weighted mean and **ordered weighted averaging (OWA)** operator being widely used in multi-criteria decision-making problems".

Researchers have known for a few years that such a type of fusing operators, with the properties of behaving like a disjunctive operator when all values are high, a conjunctive operator when all the values are low, and as a compensatory operator otherwise, do exist already, and it corresponds to the so-called family of **uninorms operators** [35].

Definition 1 A uninorm operator U is a mapping $U : [0, 1]^2 \longrightarrow [0, 1]$ having the following properties:

(a) Identity element: $\exists\, e \in [0, 1] : \forall\, x \in [0, 1],\ U(x, e) = x$
(b) Monotonicity: $U(x_1, y_1) \geq U(x_2, y_2)$ if $x_1 \geq x_2$ and $y_1 \geq y_2$
(c) Commutativity: $U(x, y) = U(y, x)$
(d) Associativity: $U(x, U(y, z)) = U(U(x, y), z)$

As presented by Appel et al. in [3]: "Uninorm operators share with t-norm and t-conorm operators the some of the properties mentioned above (commutativity, associativity and monotonicity). Furthermore, the uninorm operator generalises both the t-norm operator and the t-conorm operators. A uninorm operator has an identity element lying anywhere in the unit interval $[0, 1]$; a t-norm operator has 1 as its identity element and therefore it is a uninorm operator with identity element 1; while a t-conorm operator has 0 as its identity element and, therefore, it is a uninorm operator with identity element 0. It is well known that a uninorm operator with identity element $e \in [0, 1]$ behaves like (i) a t-norm operator when all partial IP values are below e; (ii) a t-conorm operator when all partial IP values are above e; (iii) a compensative operator in the presence of partial values below and above e. An interesting particular case of uninorm operators are the symmetric aggregative operators, i.e. uninorm operators that have a representation in terms of a single variable function. Of particular interest, the representable uninorm operator with identity element $e = 0.5$ has been characterised as the most appropriate for modelling *cardinal consistency of reciprocal preference relations*, as per Chiclana et al. [6].".

Hence, a key task would be to implement a uninorm operator to compute the intensity polarity IP of a sentence S from the IP of its associated list of sentiment-conveying particles. In our method, we have utilised the minimum operator—which as mentioned above is a type of uninorm—as follows:

$$IP(S) = \min\{IP(W_1) \ldots IP(W_n)\}. \tag{1}$$

When a sentence IP score is obtained, then the linguistic labels—or granules—$l \in G$ with highest $\mu_l(IP(S))$ is assigned to it in order to classify the sentence polarity. In the event when there exists two consecutive *granules* with the same $\mu_l(IP(S))$, then we classify the polarity of the sentence using the granule/label with a *higher meaning* as per the ordinal ordering implicitly expressed in the representation previously provided in Fig. 5. In the case when $IP(S) = 0.3$, the algorithm will assign the label *Moderate* rather than the label *Slight*, as the computed polarity.

2. Recognising sentences with a *objective/neutral* polarity: there could be sentences for which it is unclear whether they carry a positive or a negative meaning. We would classify this type of sentences as objective/neutral. In our proposed method, we consider that those sentences that get assigned the granule $G = \{Poor\}$ are prime candidates to get tagged as having *Objective/Neutral* polarity, implying that there is no opinion at all being given.

4.4.3 Enriching the Sentiment Lexicon

We are aware that challenges could become present when dealing with sentences for which there is not enough data available in our lexicon; for example, terms that are absent in the lexicon. As a consequence, our method would not be capable of generating a semantic orientation score (SOR). As a response to this type of situation, we have introduced a technique that we have called *sentiment lexicon enrichment*.

However, this latter aspect is out of the scope of this chapter. As such, the motivated reader should gain access to the work of Appel et al. [3].

4.5   Experimental Results

In this section, we will look at the results obtained during our experiments. In order to be able to establish meaningful comparisons, we will start by showing the output of other methods, namely: Naïve Bayes (NB) and Maximum Entropy (ME). Then, we will share the results achieved by applying our proposed hybrid method. We take for granted that the reader is familiar with how the aforementioned machine learning techniques operate. For more information on this subject refer to work by Alpaydin [1], Bird et al. [4], Marsland [19] and Perkins [24].

Table 5 shares the outputs obtained using the NB classifier with the test datasets accordingly specified below, while Table 6 shares the results achieved when the ME trained classifier is applied.

4.5.1   The Proposed Hybrid Method in 2-Steps (HSC & HAC)

In order to illustrate how the two steps of our method have an impact on the results, we first applied the HSC method and then the HAC component. The results of having applied the polarity determination technique are shown first followed by the output of the determination of the polarity intensity. In terms of data, our proposed hybrid method is first applied to both twitter datasets and then to the movie database set.

Table 5 Naïve Bayes classifier performance indexes

Metric	Dataset: Twitter A	Dataset: movie database
Accuracy	0.6785	0.6717
Precision	0.6315	0.6274

Table 6 Maximum Entropy classifier performance indexes

Metric	Twitter A dataset	Movie database dataset
Accuracy	0.6759	0.6757
Precision	0.6293	0.6291

HSC Results

Notice that we applied the HSC method in two passes. When running the experiments, we reset our lexicon to an initial state every time in order to simulate that we are not incorporating yet the *learnings* into the dictionaries. During the 2nd pass, we do test our method by allowing the execution of the step where the sentiment lexicon and the word dictionary have learnt new—absent or incomplete—terms.

Table 7 shows the HSC method with its 1st and 2nd passes with the Twitter A dataset. Table 8 presents the HSC 2nd pass results when applied to the Twitter B dataset. For simplicity, Table 9 displays only the HSC 2nd pass results when we utilise the Movie Review data.

It is key to highlight that the focus of our research is sentiment analysis at the sentence level. Our experiments reveal that when the data utilised contains mostly *snippets*, as in the case of Twitter examples, the better our proposed hybrid method performs. In Sect. 4.5.2, we will elaborate further on this aspect.

HAC Results

In this sub-section, we will present the results of adding the *fuzzy sets* approach into our method (HAC). As a consequence, we achieve a much *finer granularity level* into the sentiment classification process. The output of running the HAC component are shared in Tables 10 and 11.

As expected, no cases of NOSORs are found in the 2nd pass, as during the 1st pass, our proposed method **learnt** new terms that were incorporated into the senti-

Table 7 HSC classifier—Twitter A dataset performance indexes

Metric	1st pass—HSC	2nd pass—HSC
Accuracy	0.8728	0.8802
Precision	0.8280	0.8424

Table 8 HSC classifier—Twitter B dataset performance indexes

Metric	2nd pass—HSC
Accuracy	0.8655
Precision	0.8406

Table 9 HSC classifier—movie review dataset performance indexes

Metric	2nd pass—HSC
Accuracy	0.7585
Precision	0.7278

Table 10 Our classifier with increased granularity functionality (HAC)—**Positive Polarity** data set

False negatives	929
NOSOR (no semantic orientation)	35
NOSOR (2nd run)	0
True positives	4,402
Poorly granule	577
Slight granule	1,106
Moderate granule	1,041
Very granule	1,365
Most granule	313
Number of sentences total	5,331

Table 11 Our classifier with increased granularity functionality (HAC)—**Negative Polarity** data set

False positives	1,646
NOSOR (no semantic orientation)	76
NOSOR (2nd run)	0
True negatives	3,685
Poorly granule	770
Slight granule	1,089
Moderate granule	789
Very granule	864
Most granule	173
Number of sentences total	5,331

ment lexicon and then the system was capable of finding polarity scores for terms that either was initially absent or were present but incomplete. With this added granularity to our method, now we are capable of examining sentences classified in the lower end of the spectrum [0, 1] of the interval (those sentences labelled as Granule = 'Poor'), and *upgrade them* in terms of classification; these cases could be considered to be borderline with being *Objective* as opposed to *Subjective*. Examples of these could be sentences like '*The theatre was completely full*' or "*The Sinner counted with famous actors*". Those two examples are rather stating 'facts' than 'opinions'.

As we were lacking *intensity polarity annotations* we were forced to annotate 10% of all sentences in the movie review database (positive polarity). We ended up with 530 with the following distribution: 130 sentences for the 'Moderate' Granule and 100 each for the Granules 'Most', 'Very', 'Slight' and 'Poor'. Table 12 presents the metrics for this dataset. Notice that results are very good as success in predicting polarity intensity accurately was above 80% for all cases.

Table 12 Movie review dataset—HAC classifier performance

Subject	Poor	Slight	Moderate	Very	Most
Number of sentences total	100	100	130	100	100
Predicted correctly (%)	81.00	89.00	93.08	91.00	87.00
Predicted incorrectly (%)	19.00	11.00	6.92	9.00	13.00

4.5.2 Experimental Results—A Comparison

Let us now address some comparison aspects. In Table 13 we show the outputs obtained when the three methods being compared are executed against the *Twitter A dataset*—keeping in mind that the 2nd pass of HSC corresponds to the stage at which the sentiment lexicon has learnt new conveying-sentiment words and properties-. The results achieved are excellent as our HSC method exceeds by 20+% the other two techniques, namely NB and ME. In order to confirm these results, we applied another dataset, *Twitter B* to the three methods and similar outputs are produced, reassuring the researchers that the results obtained are sound.

Once we finalised the experiments using datasets containing *snippets*, then we proceed to use instead of the *movie database*, which contains more complex sentences—composed sentences and paragraphs-. The results are shared in Table 14, and immediately we notice that our HSC method continues to be the best performer, but its overall performance for the *Precision Indicator* decreases by approximately 11.5% with respect to the outputs achieved when the *Twitter datasets* were used. As we stated from the beginning of this technical report, our method has been designed to deal with snippets and the contents in the Movie Review dataset are more complex in structure and composed paragraphs are abundant. The latter challenges our HSC method to a certain extend.

Table 13 Performance indeces comparison—NB/ME versus HSC (Twitter A dataset)

Indicator	NB	ME	1st pass HSC	2nd pass HSC
Accuracy	0.6785	0.6759	0.8728	0.8802
Precision	0.6315	0.6293	0.8280	0.8424

Table 14 Performance indeces comparison—NB/ME versus HSC (movie review dataset)

Indicator	NB	ME	1st pass HSC	2nd pass HSC
Accuracy	0.6717	0.6757	0.7559	0.7585
Precision	0.6274	0.6291	0.7263	0.7278

Table 15 Impact of the incorporation of various techniques into the hybrid method—Twitter A dataset

Technique implemented	Precision	Cumulative impact (%)
Base semantic rules	76.77	
Adding effective PoS tagging	79.33	3.33
Adding smart negation handling	81.17	5.73
Adding new semantic rules (R14 & R15)	83.36	8.58
Adding semantic lexicon after it has learnt new terms (2nd pass)	84.24	9.73

Impact of Different Techniques in Hybrid Approach

Next, we attempt to show how the different techniques that made our HSC method impact the experiments' outputs as they get introduced following an orderly timeline. This way, we can appreciate how outputs change as the various techniques are added to the mix. The focus of the comparison will be the *Precision* indicator. See Table 15 for the results. Notice that the process is *cumulative* in nature as every step in the table inherits the benefits of having introduced a specific technique in the previous step. As expected, the indicator at the bottom of the table is much better, showing approximately 10 points of improvement when compared to the first step presented in the table.

Analysis of Specific Examples

Next we will present some examples of sentences that will assist us in showing some interesting aspects.

Example 1 Granule = Poor: "effective but too-tepid biopic."

Example 2 Granule = Slight: "if you sometimes like to go to the movies to have fun, wasabi is a good place to start."

Example 3 Granule = Moderate: "occasionally melodramatic, it's also extremely effective."

Example 4 Granule = Very: "the movie's ripe , enrapturing beauty will tempt those willing to probe its inscrutable mysteries."

Example 5 Granule = Most: "one of the greatest family-oriented, fantasy-adventure movies ever."

The above examples show how our method is assigning *polarity intensities* that in turn match our human 'perception' of how intense the opinions expressed in the snippets are. But what about instances where our classifier stumbles to produce good results? Next, we will present examples of sentences that were challenging for our proposed method.

Example 6 "spiderman rocks."
In this case, the classifier misunderstands the semantic of the term 'rocks' and produces a negative score.

Example 7 "it extends the writings of Jean Genet and John Rechy, the films of Fassbinder, perhaps even the nocturnal works of goya."
This sentence enumerates names of famous movie directors and performers and claims that the director of such a movie expands on the work of these well-known characters. Because our sentiment lexicon lacks 'context' about famous individuals it fails to realise that this sentence has a positive leaning.

Example 8 "after watching the movie I found myself between a rock and a hard place."
'between a rock and a hard place' can be considered an idiom, and it confuses the classifier.

4.6 Performance Comparison Against Machine Learning and State of the Art

A proper comparison cannot be achieved unless every method involved is appraised against the same datasets. Hence, the scores shared in Table 16, which follows, are as informative as possible, considering that every researcher that is reporting results was in control of the conditions surrounding the experimental arrangements. Notice that the intention behind this comparison was to establish the differences of our proposed method against *state of the art* algorithms that were fully based on machine learning techniques.

As we analyse the comparative results, we realise that our hybrid method performs 17% better than generic machine learning techniques, and 21.50% better when compared against Naïve Bayes and Maximum Entropy. Despite the fact that Socher et al.

Table 16 Comparison of results of the proposed hybrid method against state of the art algorithms

Method/Algorithm	Precision (%)
Naïve Bayes (from this report)	62.74
Machine learning as reported by Poria et al. [25]	67.35
Our method (HSC/HAC)—Movie Review dataset	72.78
Socher et al. as reported in [30]	80.00
Our method (HSC/HAC)—Twitter dataset	84.24
Socher et al. as reported in [29]	85.40
Ensemble classification as reported in Poria et al. [25]	86.21

[29, 30] have been accepted as state of the art, we can see that at least at the sentence level the proposed hybrid method performs better than Socher et al. [30] and even follows closely the performance reported in Socher et al. [29]. Without a doubt, the ensemble classification method by Poria et al. [25] shows the best results amongst all (Precision = 86.21%).

5 Conclusions

As we embarked on this research project, our hypothesis was that a hybrid system based on sentiment lexicons, semantic rules and *fuzzy sets*, should be able to perform as well as machine-learning-based techniques. In fact, our hybrid system has proven to dramatically improve the scores achieved by Naïve Bayes (NB) and Maximum Entropy (ME). When the dataset consists of *sentences*, our system does perform very well producing high scores of *accuracy* (88.02%) and *precision* (84.24%). In this case, by sentences, we mean datasets containing 'snippets' in the English language, as they are usually present in Twitter interactions. An important **enhancement** provided by our method is that the *intensity of the sentiments conveyed by the sentences* is clearly identified, extending the capability of producing *only* a polarity score (positive/negative). This aforementioned improvement is a direct result of having incorporated the *fuzzy sets module* in our proposed method.

As the amount of terms and quality of the contents of *SentiWordNet* grow, our sentiment lexicon will improve, too, bettering the results currently achieved. All in all, our research hypothesis has been proved, leading to the idea that hybrid techniques could become a key player in the advancement of the Sentiment Analysis field.

As a sub-product of our research, we have been capable as well of identifying under which conditions our proposed method under-performs. The main characteristics of those cases follow:

1. The presence of argot, idiom, jargon and lingo is very challenging for the classifier.
2. The incorporation of human-like features like 'imagery, metaphors, similes, sarcasm, humour' is difficult to deal with adequately. It is important to highlight that these language elements require knowing the context surrounding the imagery been used. This is a critical area of research that we intend to address in the short term. In the meantime, the reader is referred to the work of Justo et al. [16].
3. Further work is required to effectively manage *double-negation*, as the scope of its effect is hard to define with simple syntactic rules.

References

1. Alpaydin, E. (2010). *Introduction to machine learning* (2nd ed.). The MIT Press.
2. Appel, O., Chiclana, F., & Carter, J. (2015). Main concepts, state of the art and future research questions in sentiment analysis. *Acta Polytechnica Hungarica - Journal of Applied Sciences, 12*(3), 87–108.
3. Appel, O., Chiclana, F., Carter, J., & Fujita, H. (2016). A hybrid approach to the sentiment analysis problem at the sentence level. *Knowledge-Based Systems, 108*, 110–124. New Avenues in Knowledge Bases for Natural Language Processing.
4. Bird, S., Loper, E., & Klein, E. (2009). *Natural language processing with python*. O'Reilly Media Inc.
5. Cambria, E., Olsher, D., & Rajagopal, D. (2014). Senticnet 3: A common and common-sense knowledge base for cognition-driven sentiment analysis. In *Proceedings of the Twenty-Eighth AAAI Conference on Artificial Intelligence*, July 27–31, 2014, Québec City, Québec, Canada (pp. 1515–1521).
6. Chiclana, F., Herrera-Viedma, E., Alonso, S., & Herrera, F. (2009). Cardinal consistency of reciprocal preference relations: A characterization of multiplicative transitivity. *IEEE Transactions on Fuzzy Systems, 17*(1), 14–23.
7. Das, S. R., Chen, M. Y., Agarwal, T. V., Brooks, C., Chan, Y. S., Gibson, D., et al. (2001). Yahoo! for Amazon: Sentiment extraction from small talk on the web. In *8th Asia Pacific Finance Association Annual Conference*.
8. Dzogang, F., Lesot, M.-J., Rifqi, M., & Bouchon-Meunier, B. (2010). Expressions of graduality for sentiments analysis—A survey. In *2010 IEEE International Conference on Fuzzy Systems (FUZZ)* (pp. 1–7).
9. Esuli, A., & Sebastiani, F. (2005). Determining the semantic orientation of terms through gloss classification. In *Proceedings of the 14th ACM International Conference on Information and Knowledge Management, CIKM '05*, New York, NY (pp. 617–624). ACM.
10. Esuli, A., & Sebastiani, F. (2006). SentiWordNet—A publicly available lexical resource for opinion mining. In *Proceedings of the 5th Conference on Language Resources and Evaluation (LREC06)* (pp. 417–422).
11. Esuli, A., & Sebastiani, F. (2006). *SentiWordNet: A high-coverage lexical resource for opinion mining*. Technical report ISTI-PP-002/2007, Institute of Information Science and Technologies (ISTI) of the Italian National Research Council (CNR).
12. Hajmohammadi, M. S., Ibrahim, R., Selamat, A., & Fujita, H. (2015). Combination of active learning and self-training for cross-lingual sentiment classification with density analysis of unlabelled samples. *Information Sciences, 317*(C), 67–77.
13. Hatzivassiloglou, V., & McKeown, K. (1993). Towards the automatic identification of adjectival scales: Clustering adjectives according to meaning. In L. K. Schubert (Ed.), *ACL: 31st Annual Meeting of the Association for Computational Linguistics (ACL)*, 22–26 June 1993, Ohio State University, Columbus, Ohio, USA, Proceedings (pp. 172–182). ACL.
14. Hatzivassiloglou, V., & McKeown, K. R. (1997). Predicting the semantic orientation of adjectives. In *Proceedings of the 35th Annual Meeting of the Association for Computational Linguistics (ACL) and the 8th Conference of the European Chapter of the ACL*, New Brunswick, NJ, USA (pp. 174–181). ACL.
15. Hu, M., & Liu, B. (2004). Mining and summarizing customer reviews. In: *Proceedings—ACM SIGKDD International Conference on Knowledge Discovery and Data Mining (KDD-2004 full paper)*, Seattle, Washington, USA, 22–25 Aug 2004.
16. Justo, R., Corcoran, T., Lukin, S. M., Walker, M. A., & Torres, M. I. (2014). Extracting relevant knowledge for the detection of sarcasm and nastiness in the social web. *Knowledge-Based Systems, 69*, 124–133.
17. Kamps, J., Marx, M., Mokken, R. J., & de Rijke, M. (2004). Using WordNet to measure semantic orientations of adjectives. In *Proceedings of LREC-04, 4th International Conference on Language Resources and Evaluation*, Vol. IV of *LREC '04* (pp. 1115–1118).

18. Liu, B. (2012). *Sentiment analysis and opinion mining*. Synthesis lectures on human language technologies (1st ed.). Morgan and Claypool Publishers.
19. Marsland, S. (2009). *Machine learning—An algorithmic perspective*. Chapman & Hall/CRC machine learning & pattern recognition series (1st ed.). CRC Press.
20. Miller, G. A. (1956). The magical number seven, plus or minus two: Some limits on our capacity for processing information. *Psychological Review, 63*, 81–97.
21. Nadali, S., Murad, M., & Kadir, R. (2010). Sentiment classification of customer reviews based on fuzzy logic. In *2010 International Symposium in Information Technology (ITSim)*, Kuala Lumpur, Malaysia, June 2010 (Vol. 2, pp. 1037–1040).
22. Pang, B., & Lee, L. (2005). Seeing stars: Exploiting class relationships for sentiment categorization with respect to rating scales. In *Proceedings of the 43rd Annual Meeting on Association for Computational Linguistics (ACL '05)* (pp. 115–124).
23. Pang, B., Lee, L., & Vaithyanathan, S. (2002). Thumbs up? Sentiment classification using machine learning techniques. In *Proceedings of the Association for Computational Linguistics (ACL-02) Conference on Empirical Methods in Natural Language Processing (EMNLP)* (Vol. 10, pp. 79–86).
24. Perkins, J. (2010). *Python text processing with NLTK 2.0 cookbook*. Packt Publishing.
25. Poria, S., Cambria, E., Winterstein, G., & Huang, G.-B. (2014). Sentic patterns: Dependency-based rules for concept-level sentiment analysis. *Knowledge-Based Systems, 69*, 45–63.
26. Potts, C. (2011). *Sentiment Symposium Tutorial: Linguistic Structure (part of the Sentiment Analysis Symposium held at San Francisco, November 8–9, 2011)*. Stanford Department of Linguistics, Stanford University. Accessed date: December 2014, November 2011.
27. Ravi, K. R., & Ravi, V. (2015). A survey on opinion mining and sentiment analysis: Tasks, approaches and applications. *Knowledge-Based Systems, 89*(C), 14–46.
28. Sadegh, M., & Othman, R. I. Z. A. (2012). Combining lexicon-based and learning-based methods for Twitter sentiment analysis. *International Journal of Computers & Technology, 2*(3), 171–178.
29. Socher, R., Huval, B., Manning, C. D., & Ng, A. Y. (2012). Semantic compositionality through recursive matrix-vector spaces. In *Proceedings of the 2012 Joint Conference on Empirical Methods in Natural Language Processing and Computational Natural Language Learning, EMNLP-CoNLL 2012, July 12–14, 2012, Jeju Island, Korea* (pp. 1201–1211).
30. Socher, R., Perelygin, A., Wu, J. Y., Chuang, J., Manning, C. D., Ng, A. Y., et al. (2013). Recursive deep models for semantic compositionality over a sentiment treebank. In *Proceedings of the Conference on Empirical Methods in Natural Language Processing (EMNLP)* (p. 1631).
31. Subasic, P., & Huettner, A. (2001, August). Affect analysis of text using fuzzy semantic typing. *IEEE Transactions on Fuzzy Systems, 9*(4), 483–496.
32. Wang, S., & Manning, C. D. (2012). Baselines and bigrams: Simple, good sentiment and topic classification. In *Proceedings of the 50th Annual Meeting of the Association for Computational Linguistics (ACL 2012): Short Papers* (Vol. 2, pp. 90–94).
33. Wiebe, J. (2000). Learning subjective adjectives from corpora. In *Proceedings of the Seventeenth National Conference on Artificial Intelligence and Twelfth Conference on Innovative Applications of Artificial Intelligence* (pp. 735–740). AAAI Press.
34. Xie, Y., Chen, Z., Zhang, K., Cheng, Yu., Honbo, D. K., Agrawal, A., & Choudhary, A. N. (2014). Mu SES: A multilingual sentiment elicitation system for social media data. *IEEE Intelligent Systems, 29*(4), 34–42. July.
35. Yager, R. R., & Rybalov, A. (1996). Uninorm aggregation operators. *Fuzzy Sets and Systems, 80*(1), 111–120. Fuzzy Modeling.
36. Zadeh, L. A. (1973). Outline of a new approach to the analysis of complex systems and decision processes. *IEEE Transactions on Systems, Man and Cybernetics, Part B-Cybernetics, SMC-3*, 28–44.
37. Zadeh, L. A. (2001). A new direction in AI: Toward a computational theory of perceptions. *AI Magazine, 22*(1), 73–84.
38. Zadeh, L. A. (2002). From computing with numbers to computing with words—From manipulation of measurements to manipulation of perceptions. *International Journal of Applied Mathematics and Computer Science (AMCS), 12*(3), 307–324.

Consensus in Sentiment Analysis

Orestes Appel, Francisco Chiclana, Jennifer Carter, and Hamido Fujita

Abstract The objective of this chapter is to present a method applicable in group decision-making where computing the opinion of the majority of participants is key. In this article, we present a method that makes use of Induced Ordered Weighted Averaging (IOWA) operators to aggregate a majority opinion out of a number of Sentiment Analysis (SA) classification systems. The numerical output of each SA classification method is used as input to a carefully chosen IOWA operator that is semantically equivalent to the fuzzy linguistic quantifier 'most of'. The object of the aggregation will be the intensity of the previously determined sentence polarity in such a way that the results represent what the majority thinks.

Keywords Sentiment analysis · Consensus · Majority support · Sentiment aggregation · Ordered weighted averaging · OWA · Induced ordered weighted averaging

1 Sentiment Aggregation by Consensus

What could we do in situations where the majority's opinion is essential in a given decision-making process? There could be many situations representative of this, like experts finding common ground on diagnosing a potential problem in an X-ray, law-makers looking for fairness in a final decision or high-school teachers deciding as

O. Appel (✉) · F. Chiclana
Institute of Artificial Intelligence (IAI), De Montfort University, Leicester, UK
e-mail: Orestes.Appel@email.dmu.ac.uk

F. Chiclana
e-mail: chiclana@dmu.ac.uk

J. Carter
University of Huddersfield, Huddersfield, UK
e-mail: j.carter@hud.ac.uk

H. Fujita
Intelligent Software Systems Laboratory, Iwate Prefectural University, Takizawa, Iwate, Japan
e-mail: issam@iwate-pu.ac.jp

© Springer Nature Switzerland AG 2021
J. Carter et al. (eds.), *Fuzzy Logic*,
https://doi.org/10.1007/978-3-030-66474-9_3

a group on the recipients of scholarships. In this book's chapter, we will present a technique that utilises the so-called Induced Ordered Weighted Averaging (IOWA) in order to generate *consensus* among participants. Given the abundance of Sentiment Analysis (SA) methods available today, we will apply the aforementioned *Consensus Technique* to combine the outputs—the latter in the unit interval-of three SA methods and then achieve an outcome that corresponds to a semantic leading towards what the majority thinks. Hence, the numerical outcomes of these three SA classification systems are used as input to an IOWA-based *Consensus Operator* which is semantically analogous to the fuzzy linguistic quantifier 'most of' devised by Yager [15, 16]. During our experiments, the *Consensus Operator* achieved far better results than other commonly accepted techniques.

2 Introduction

In the case of the effort presented in this chapter, the participating SA techniques would correspond to any number of SA classification techniques (*n* of them, with $n \geq 2$). Our experiments have been performed using three specific SA classification systems: (i) a Hybrid Approach to SA implemented by Appel et al. [1], (ii) a Naïve Bayes system [10] and (iii) Maximum Entropy [3]. The main idea is that a number of classification systems will perform separately their classification tasks, and then the results would be processed by our proposed Consensus Operator. Let us keep in mind that we need to make sure that the value we would obtain is actually equivalent to the majority. We are searching for a majority-driven aggregation mechanism that stresses some outputs over others among the set being considered. As a consequence, we propose using an IOWA-based operator [7, 18] to aggregate the outcome of a number of classification systems utilising an induced guiding principle. We will discuss further Yager's creation—the Induced Ordered Weighted Averaging (OWA) operator [16]—and will provide our reasoning for selecting it as the aggregation mechanism of choice.

3 Fuzzy Majority in Collective Decision Making Modelled with an IOWA Operator

It has been already established by Yager [14, 18] that the OWA operator provides a *parameterized* family of mean type aggregation operators. The parameterized component is directly associated with the utilised weighting vector. In this section, we will describe in more detail OWA operators and fuzzy majority concepts.

3.1 The Linguistic Quantifier in Fuzzy Logic

A linguistic quantifier generalises the idea of quantification of classical logic—the same way as other fuzzy logic concepts relate to classical logic-. In the latter there exist two types of quantifiers that can be used in propositions: the universal quantifier and the existential quantifier. As per Pasi and Yager [11], through the utilisation of linguistic quantifiers we are capable of referencing a variable number of elements of the domain of discourse. This referencing can be achieved in a *crisp* fashion or in a *fuzzy* way. See Table 1 for details. Pasi and Yager [11] differentiate between two types of fuzzy quantified propositions as presented in Table 2. According to Zadeh [19], in fuzzy logic domain, the quantifiers have been defined as fuzzy subsets of two main types: absolute and proportional. Quoting Pasi and Yager [11], "absolute quantifiers, such as *about* 7, *almost* 6, etc. are defined as fuzzy subsets with membership function $\mu_Q : \Re^+ \rightarrow [0, 1]$, where $\forall x \in \Re^+$; $\mu_Q(x)$ indicates the degree to which the amount x satisfies the concept Q. In addition, as per [11], proportional quantifiers like *most*, or *about* 70%, are defined as fuzzy subsets of the unit interval: $\mu_Q : [0, 1] \rightarrow [0, 1]$, where $\forall x \in [0, 1]$, $\mu_Q(x)$ indicates the degree to which the proportion x satisfies the concept Q". In the rest of this article, we will use Q instead of μ_Q and $Q(x)$ in place of $\mu_Q(x)$ in an effort to simplify the used expressions.

Table 1 Types of referencing to participants in the domain of discourse

Reference type chosen	Cases
Crisp type	*half* of the participants, *all* of the participants, *at least k* of the participants
Fuzzy type	*most* of the participants, *approximately k* of the participants, *some* of the participants

Table 2 Types of fuzzy quantified propositions

Type of fuzzy quantified proposition	Proposition constituents	Proposition statement	Some examples
"Q X are Y"	Q is a Linguistic quantifier, Y is a fuzzy predicate, X is a set of elements	Q participants of set X satisfy the fuzzy predicate Y	Most of the criteria are satisfied by alternative A_i, in which Q Is *most*, X is the set of the *criteria*, and Y *satisfies alternative A_i*
"Q B X are Y"	Q is a Linguistic quantifier, B *and* Y are fuzzy predicates, X is a set of elements	Q participants of set X which satisfy the fuzzy predicate B, also satisfy the fuzzy predicate Y	Most of the important criteria are satisfied by alternative A_i, in which B Is *important*

3.2   Linguistic Quantifiers as Soft Specifications of Majority-Based Aggregation

The central point of this section will be elaborating on linguistic quantifiers that are monotonic and non-decreasing. Examples of those are *most* and *at least*. As we are looking for a quantifier related to the concept of majority, we will devote our attention to the quantifier "most". As discussed by Pasi and Yager [11], our objective is to apply linguistic quantifiers "in guiding an aggregation process aimed at computing a value that synthesises the majority of values to be aggregated. This aimed value is known as a 'majority opinion'". Let us consider a decision-making process that is multi-agent in nature. The aggregation of opinion/thoughts of the majority is a fundamental idea. OWA operators can be built on top of the concept of the fuzzy definition of a linguistic quantifier. As mentioned in [11], the researchers discuss if the result of aggregating a set of values using a quantifier that conveys the idea of *majority*, will be at the same time equivalent to *the majority of values*. The semantics pushed-in during the aggregation is essential to be able to truly reflect the concept of *majority*. Two options of possible OWA semantics that have been discussed in [11] are: (i) OWA operators as an aggregation guided by 'majority' linguistic quantifiers, and (ii) Induced OWA operators as drivers of a *majority opinion*.

1. Case i—"The semantic of OWA operators is an aggregation guided by 'majority' linguistic quantifiers" [11]: Pasi and Yager see the OWA operator as an aggregation operator taking as argument a number of values and returning a number in the unit interval. As we know, the weights of the OWA operator will determine the behaviour of the aggregation operator.

   - One possible semantics considered is that one that presents the OWA operator as a generalisation of the concept of a summarising operator, for example, $opr_i = 1/n$ for all i produces a simple average as all components in the aggregation obtained do in fact contribute *all the same* to the final result.
   - A second possible semantics for the Ordered Weighted Averaging operator to be considered, is the one in which the operator becomes a generalisation of the *there exists* and *for all* classical logic quantifiers.

   In [11], Pasi and Yager claim that "these semantics of the aggregation do *not* really reflect the concept of majority in group decision-making applications". Hence, the authors have chosen the semantics presented below in Case (ii).

2. Case ii—Using Induced OWA operators to achieve a *majority opinion*: this aligns with the concept of majority as it is more often used in group decision-making applications, where more than one opinion provider participates, and it is truly closer to the linguistic quantifier *most*, as we aim to obtain a meaning that represents a majority of the involved experts sharing a common opinion. In this case, what we pretend to say by *majority* is *most* of them. As a matter of fact we are researching for an operator that computes 'an average-like aggregation of "a **majority** of values that are similar"'.

   Pasi and Yager [11] propose such an aggregation that according to them *does have* a majority semantics. Their proposal is based on the utilisation of OWA operators that include "an inducing ordering variable which is based on a proximity metric over the elements to be aggregated" [11]: the so-called IOWA operator. The authors focus on a method for calculating the weights used in the OWA operator that would allow them to obtain the weights from a functional form "$Q : [0, 1] \rightarrow [0, 1]$ such that $Q(0) = 0$, $Q(1) = 1$, and $Q(x) \geq Q(y)$ for $x > y$ corresponding to a fuzzy set representation of a proportional monotone qualifier. For a given value $x \in [0, 1]$, the $Q(x)$ *is the degree to which* x *satisfies the fuzzy concept being represented by the quantifier*" [11]. Based on this function Q, the OWA vector is computed in the manner described below in 1 [11]:

$$opr_i = Q(i/n) - Q((i - 1)/n) \tag{1}$$

Hence, opr_i corresponds to "the increase of the satisfaction in getting i with respect to $(i - 1)$ criteria satisfied". The authors claim that, for example, "if they were going to define the so-called (weighting) vector of the OWA operator that is associated to the linguistic quantifier **most**, a possible *membership function* of the **most** quantifier would be":

$$\mu_{most}(x) = \begin{cases} 1 & \text{if } x \geq 0.9 \\ 2x - 0.8 & \text{if } 0.4 < x < 0.9 \\ 0 & \text{if } x \leq 0.4 \end{cases} \tag{2}$$

The main idea being exploited here is that "most similar values must have close positions in the induced ordering in order to appropriately be aggregated" [11]. We interpret this by meaning that *similar values should be closer to each other in the support vector used in the operator and the final output should certainly reflect in a stronger fashion the sentiments of the majority.* "To this aim, our intent is to take the most similar values in the quantity specified by the quantifier and apply to them an averaging operator." [11]. What is truly needed is to create the capability to compute the similarities among the opinion values being considered. "The values of the inducing variable of the IOWA operator are obtained by means of a function of the *similarities* between *pairs* of the opinion values." [11]. Such a function is defined using a support function model introduced by Ronald Yager in [17].

As it is vital to our proposed method, we reproduce below the discussion presented by Yager [17], in order to obtain the aforementioned support function.

A support function $Sfun$, is a two-valued function that calculates a value $Sfun(a, b)$ which expresses the support from b for a, where α is a desired tolerance. "The more similar, the more close two values are, the more they support each other". The more elevated the tolerance is, the less we impose that the values a and b have to be closer to each other.

$$Sfun(x_i, x_j) = \begin{cases} 1 & \text{if } |x_i - x_j| < \alpha \\ 0 & \text{otherwise} \end{cases} \qquad (3)$$

If we were to aggregate a set of values and we wanted to order them in an increasing fashion of support, "we compute for each value the sum of its support values with respect to all others values to be aggregated" [11]. Then, for each decision-maker opinion "we sum all the supports it has in order to obtain its overall support". These overall supports for a decision-maker's opinion are utilised as "the values of the order-inducing variable". Below we show an example created by Yager himself, with the purpose of clarifying the concept of a support vector.

Let us assume that the threshold parameter is $\alpha = 0.4$ and that we encounter the following values that require aggregation:

$$x_1 = 0.9, \quad x_2 = 0.7, \quad x_3 = 0.6, \quad x_4 = 0.1, \quad x_5 = 0$$

Using Eq. 3 above, we obtain the following support values:

$$
\begin{array}{llll}
Sfun(x_1, x_2) = 1 & Sfun(x_1, x_3) = 1 & Sfun(x_1, x_4) = 0 & Sfun(x_1, x_5) = 0 \\
Sfun(x_2, x_1) = 1 & Sfun(x_2, x_3) = 1 & Sfun(x_2, x_4) = 0 & Sfun(x_2, x_5) = 0 \\
Sfun(x_3, x_1) = 1 & Sfun(x_3, x_2) = 1 & Sfun(x_3, x_4) = 0 & Sfun(x_3, x_5) = 0 \\
Sfun(x_4, x_1) = 0 & Sfun(x_4, x_2) = 0 & Sfun(x_4, x_3) = 0 & Sfun(x_4, x_5) = 1 \\
Sfun(x_5, x_1) = 0 & Sfun(x_5, x_2) = 0 & Sfun(x_5, x_3) = 0 & Sfun(x_5, x_4) = 1
\end{array}
$$

The overall support for each x_i is computed by adding the support values for x_i. The support for each x_i is denoted as $supp_i$:

$$supp_1 = 2, \quad supp_2 = 2, \quad supp_3 = 2, \quad supp_4 = 1, \quad supp_5 = 1$$

Pasi and Yager [11] claim that it becomes evident that there are two clusters of similar values in s_i. Hence, the support function (Eq. 3) induces "a clustering of the arguments which can be controlled by the choice of the threshold parameter α in the aforementioned function $Sfun(x_i, x_j)$". In the above example we can see that there are two clusters, 2's and 1's, with some ties of the support values. Yager claims that in order to address the ties "we could impose a 'stricter' condition by setting $\alpha = 0.3$". Then, the new support vector would look like:

$$supp_1 = 1, \quad supp_2 = 2, \quad supp_3 = 1, \quad supp_4 = 1, \quad supp_5 = 1$$

This result enables us to "order the elements to be aggregated in the following increasing order of similarity" [11]:

$$Induced\ Similarity\ Order,\ I = [0 \quad 0.1 \quad 0.6 \quad 0.9 \quad 0.7]$$

Pasi and Yager conclude that the use of an adequate support function "enables us to induce an ordering based on proximity". This concept is paramount in understanding IOWA operators, as now it would be possible to us to generate a *majority-based aggregation* of the previous values a_i. As per Yager, "The selected IOWA operator should then correspond to the linguistic quantifier *most*. Let us recall the definition of the linguistic quantifier *most* presented in Eq. 2". This linguistic quantifier when used in Eq. 1 would derive the weighting vector $W = [0 \ \ 0 \ \ 0.4 \ \ 0.4 \ \ 0.2]$. Aggregating the vector I we obtain: $IW = 0.74$. However, it is discouraging that the 5th element of the vector W is smaller than the 4th element, and despite the fact that this condition is coherent with "the interpretation of the weights as increase in satisfaction in having $i + 1$ with respect to having i criteria satisfied", the expectation is that "in an aggregation with semantics of *majority* what would be expected is that the weights of the weighting vector are non-decreasing.". In fact, as in the induced order of the arguments the top value is the 'most supported' one from all the other values (the most representative) "it should be more emphasised than the others, or at least not less emphasised.". Pasi and Yager argue that a new strategy is required for the construction of the weighting vector that would contribute to generate a value more representative of a majority of the aggregated elements. The objective of this new strategy is to stress the most supported values in the resulting aggregation, i.e. the values shown on the right-hand side of the vector of values participating in the aggregation do have more *influence* in the aggregation. As such, Pasi and Yager propose the following process for the construction of a weighting vector with non-decreasing weights.

In order to calculate the non-decreasing weights of the weighting vector, the authors define the values $t_1, t_2, \ldots, t_n$ based on a modification of the $supp_1, supp_2, \ldots, supp_n$ values:

$$t_i = supp_i + 1. \tag{4}$$

In [11] the authors claim that by doing this manipulation, "the similarity of the value a_i with itself (similarity value equal to 1) is also included in the definition of the overall support for a_i. The t_i values are in increasing order, that is t_1 is the smallest value among the t_i. On the basis of the t_j values, the weights of the weighting vector are computed as follows":

$$w_i = \frac{Q(t_i/n)}{\sum_{i=1,\ldots,n} Q(t_i/n)} \tag{5}$$

"The value $Q(t_i/n)$ denotes the degree to which a given member of the considered set of values represents the *majority*". As such, Eq. 5 is the weights semantic we will apply to our aggregation problem (Fig. 1).

Our Research Objective: as we intend to find consensus driven by a support-based majority mechanism, we are attempting to *replace a number n of human agents in*

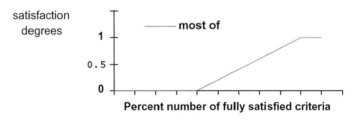

Fig. 1 A possible definition of the linguistic quantifier *most*, as presented in [11], page 395

a collective decision-making process with the output of a number n of classification system and aggregate these outputs with a method that semantically represents the concept of a majority opinion (for n ≥ 2).

4 The Proposed IOWA Approach to Sentiment Analysis

In this section we will describe how IOWA operators could be used to implement a fuzzy majority approach in the presence of recommendations (outputs) supplied by a number of classification systems. Figure 2 depicts graphically such an aggregation approach. Below we will provide more details about our proposed IOWA aggregation operator that will address the consensus of opinion problem in sentiment analysis.

4.1 The Concept of Fuzzy Majority Implemented Using IOWA Operators

"Constructing a *majority opinion* could be explained as the collective evaluation of a majority of the agents involved in the decision problem" [11]. The authors mentioned in this section provide information about the Ordered Weight Averaging operators and some of its applications: Pérez et al. [12] and León et al. [8], Yager [16], Chiclana et al. [6], Pasi and Yager [11]. In addition, Bordogna and Sterlacchini [4], Boroushaki and Malczewski [5] supply very interesting examples of real-life applications of Ordered Weighted Averaging operators.

4.2 Fuzzy Majority in Determining Intensity of the Polarity of Predetermined Subjectivity

Let us face the challenge of determining the sentiment carried by a given snippet S_k using the recommendations generated by several systems. Once every one of the participant systems performed their computations, we are interested in knowing

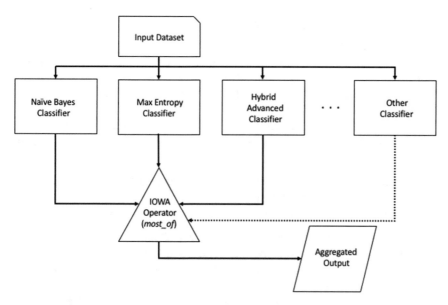

Fig. 2 IOWA$_{most}$ operator aggregating classifier methods outputs

what constitutes the opinion/thoughts of the majority. Basically, we would like to *aggregate sentiments associated with the polarity* value of snippet S_k previously obtained by using several SA classification methods. The final output will be the 'induced aggregation of the majority' with regard to sentence S_k when all the different contributions of all the participating techniques are taking into account. All the considered methods will produce their individual 'opinion' with regard to whether a sentence carries either a positive or a negative implication. For the sake of the argument we will name these methods $\{MT_1, MT_2 \ldots MT_n\}$. In the experiments realised, these techniques are the Naïve Bayes technique, the Maximum Entropy Classification System and a Hybrid Classification technique devised by the authors [1]. Figure 2 displays a graphical representation of the manner the Induced Ordered Weighted Averaging operator, using the specific semantic 'most', is applied in this practical problem. During the experimental phase, we will be using the outputs of three classification methods: (i) Maximum Entropy, (ii) Naïve Bayes and (iii) the Hybrid Method classifier designed and implemented by the authors. As we will show, our IOWA$_{most}$ operator is capable of taking as input *any* number of outputs belonging to a number of methods, with no theoretical limit to the number of methods' outputs that could be used. Pragmatically speaking, an aggregation of at least two methods is required. Hence, the condition $n \geq 2$ is enforced.

4.3   Experimental Results Obtained

In order to afford to do a fair comparison, we will measure how the IOWA_{most} operator does when fared with the *Mean* and *Median* techniques [13]. Before we share the experimental results, we will describe the methodology as well as and the datasets used during these experiments.

As two of the participating classifiers are supervised machine learning systems, annotated data is required. The data used was not initially annotated, so the annotating process is a pre-requisite.

1. Assign intensity labels in $G = \{Poor; Slight; Moderate; Very; Most\}$ to the randomly selected 500 sentences.
2. Devise a criteria to discern the concept of what *consensus* would look like.

Both tasks above were executed by qualified individuals, data scientists and a linguist. For item number one above, the three individuals assigned a label to each of the 500 sentences according to their own criteria. The English-major resolved conflicts as they appeared, as a final decision on the assigned intensity polarity label was made. For item number two, the approach followed was to observe how the combined classification scores were fused together, noticing whether the tested operators were capable to compensate for extreme values, close values and normally distributed occurrences. The expected effect of the IOWA Operator is precisely to *compensate* for the existence of outliers and to attempt to produce a number that reflects closely the semantic of the quantifier being used. In this case, it was the linguistic quantifier *most* that is driving the aggregation making it converge towards the opinion of the majority. As such, when the resulting aggregation was analysed, if the resulting number looked like one that had been obtained by *smart* aggregation/compensation, the score was considered to represent a successful case of a consensus-type aggregation.

4.4   Datasets Used

The datasets used in the experiments were initially released by Pang and Lee [9]. This dataset is known as well as the Movie Review Dataset, available at: http://www.cs.cornell.edu/people/pabo/movie-review-data/. Before we can actually use the 'answers' provided by of all the classifiers utilised as an input to our *Consensus Operator* based on the IOWA_{most} quantifier, we must perform a normalisation process. Hence, all scores participating in this process have been converted to values that are part of the interval $[0, 1] \in \mathbb{R}$, where S_k corresponds to any given snippet in the test dataset and $mt_i = \{mt_1, mt_2, \ldots, mt_n\}$ represents the SA systems i being combined (with $n \geq 2$), then:

$$\text{IOWA}_{most}^{S_k}(mt_1, mt_2, \ldots, mt_n) = \Theta^{S_k} \tag{6}$$

Once the aggregation with the semantic equivalent the *opinion of the majority* has been computed, then we should find out to which intensity level that given value Θ corresponds to. For that specific purpose, we will apply the classification method mentioned in [2], which is—for completeness—partially reproduced below. We have used trapezoidal membership functions described by the following 4-tuple (a, b, c, d):

$$
\mu_{\tilde{A}}(x) = \begin{cases} 0 & \text{if } x \leq a; \\ \frac{x-a}{b-a} & \text{if } a \leq x \leq b; \\ 1 & \text{if } b \leq x \leq c; \\ \frac{d-x}{d-c} & \text{if } c \leq x \leq d; \\ 0 & \text{if } d \leq x. \end{cases} \tag{7}
$$

The following granules—a concept introduced by Zadeh in [20]—ranging over the perception of the *intensity of the polarity* (*positivity* or *negativity*) of a given sentence S are suggested: $G = \{Poor; \; Slight; \; Moderate; \; Very; \; Most\}$, with the following 4-tuples *membership functions*, MF:

- MFun (Poor): (0, 0, 0.050, 0.150)
- MFun (Slight): (0.050, 0.150, 0.250, 0.350)
- MFun (Moderate): (0.250, 0.350, 0.650, 0.750)
- MFun (Very): (0.650, 0.750, 0.850, 0.950)
- MFun (Most): (0.850, 0.950, 1, 1)

The aggregated value Θ previously presented in Eq. 6 will take on the value x in Eq. 7 and in consequence, a proper linguistic label $\in G$ will be generated. This value is associated to the polarity intensity of a sentence S_k ($\mu_{\tilde{A}}(\Theta^{S_k}) \in G$—in essence, how negative or how positive a sentence may be.

4.5 Utilised Comparison Criteria

We are attempting to figure out which method is semantically nearer to "the opinion of the majority" among the tested classification systems. The datasets available in the movie review database include 5,331 sentences. For this exercise, we have only annotated 500 of them, which represents approximately 10% of the total universe. We have assigned to each of the 500 sentences a value $val_k \in G$, which has been carefully estimated by an exhaustive analysis of the SA polarity outcomes of the three classifying systems already mentioned.

4.6 Non-OWA Aggregation—The Outputs of the Three Classification Methods Combined Without the Application of the IOWA Operator

As a baseline, we have combined directly the outputs of the three chosen SA methods using *average* and *median*. The outcomes are summarised below. Table 3 shows the performance associated to indexes for the *Mean* and *Median* techniques.

4.7 OWA Aggregation Using Operator IOWA$_{most}$

The results of using IOWA$_{most}$ are shown in Table 4 whilst Table 5 presents the results of all methods tested.

The proposed *Consensus Operator* driven by the IOWA operator yields results that fully match our perception of what the majority would be, and as a consequence it is a much better option—by large—when compared to the other two options that have been proposed. It is important to observe that the core difference amidst the results attained by exercising two different tolerance values (0.5 and 0.3, respectively) is that when IOWA$_{most}$ is executed, another *linguistic label* in the fuzzy set G will be associated to the processed sentence. As such, the value computed for a given

Table 3 *Arithmetic Mean* and *Median* performance indexes

Classification	Method	Value
It is representative of the opinion of the majority	Mean	388
It is *not* representative of opinion of the majority	Mean	112
% of occurrences meaning success	Mean	77.60
It does represent the opinion of the majority	Median	337
It does *not* represent the opinion of the majority	Median	163
% occurrences of success	Median	67.40

Table 4 IOWA$_{most}$ operator for both tolerance $= 0.30$ and 0.50

It does convey the opinion of the majority	500
It does **not** convey opinion of the majority	0
% of success	100.00

Table 5 Performance indexes compared

Classification method	Median	Arithmetic mean	**IOWA**$_{most}$
% of success	67.40%	77.60%	100.00%

a sentence previously labelled as 'Moderate' with $\alpha = 0.3$ could now slide gently towards a different granule in G once the tolerance value α is set to 0.5. It is key to notice that "the lower the *tolerance* value, the more demanding and strict the IOWA operator is on how closely the values in the aggregation are supposed to support each other".

4.8 Specific Examples of Applying the IOWA$_{most}$ Operator

In this section, we will share some examples of having applied the IOWA$_{most}$ operator to specific sentences. The chosen examples display the outcome values of three selected classification methods. Those methods are: Naïve Bayes, Maximum Entropy and a hybrid method designed by the authors. We will call those classification systems, c_1, c_2, c_3, with each returning a number in the interval $[0, 1]$.

Let us discuss the results shown in Table 6. In the case of Example No. 1, the *Consensus Algorithm* used a $Tolerance = 0.5$. This parameter does not call for a strong support to each other by the elements participating in the aggregation. As a consequence, by using a rather *neutral* value for tolerance, the *Consensus algorithm* generates an output that is very similar to the one achieved by the *Mean*. However, when in Example No. 2 the tolerance is set to $Tolerance = 0.3$, we can see that the *Consensus System* calls for a much tighter support by each of the values to be aggregated. The result shows that the values of the first two methods c_1 and c_2, which happen to be representative of the way the majority feels, is nearer to the elements with a higher support value: 0.5643 and 0.5089. As a consequence, the output produced by the Consensus Method is not that near to the *Mean*, nor the *Median*. Once we examine Example No. 3, which again utilises $Tolerance = 0.3$, we notice that the *Consensus Method* outputs a value in the range $(0.6000, 0.6825)$ that by supporting each other in a stronger way, attains an output that represents the majority's opinion.

Table 6 Three examples—aggregation of $[c_1, c_2, c_3]$ classification systems

Example No.	Method 1	Method 2	Method 3	Mean	Median	Consensus	α
1	0.9591	0.5000	1.0000	0.8197	0.9591	0.8197	0.5
2	0.5646	0.5089	1.0000	0.6911	0.5643	0.5367	0.3
3	0.9895	0.6825	0.6000	0.7573	0.6825	0.6412	0.3

4.9   The Tolerance Value α and Its Role

If we follow the description provided in Sect. 3.2, Eq. 3 for the Induced Ordered Weighted Averaging operator, which has become the foundation for our *Consensus Operator*, we can appreciate that the support function $Sfun(x_i, x_j)$ is computed by using a tolerance value α. This latter value will have a significant importance in the aggregation that is generated.

The distance between x_i and x_j will determine the results of the computation of $|x_i - x_j|$. The more separated x_i and x_j are, the larger the difference between these two elements will become, potentially leading to the coefficient α having to be elevated, in order for the support function $Sfun$ being capable of outproducing 1 instead of 0. If the value of α decreases then the less we would be imposing a condition of larger proximity between x_i and x_j. In the same line of thought, in the case when the difference of x_i and x_j is larger than α then the enforcement of the participation of these values in the full aggregation would be minimal ($t_i = 1$ in Eq. 4 as $supp_i$ becomes 0). As α goes up in value, the output of the *Consensus Operator* goes up as well, pushing the result of the aggregation towards 1, assigning then a label-value on the extreme right of G (more positive in terms of the sentiment expressed by the group). Further analysis of the right configuration of the tolerance coefficient α is required, so we can better define the "strength" of the consensus being computed.

5   Conclusion

The method we have introduced in this chapter enables us to compute the opinion of the majority by using an Induced OWA operator (IOWA) that reflects the semantics of the fuzzy quantifier *most*. By doing this, we have created a *Consensus Operator*. The experimental results presented above demonstrate that the proposed *aggregation by consensus* method is a much better option than the two other approaches shown, provided that the objective is to obtain an opinion that represents the opinion or sentiment of the majority.

Further research is still required, but we are persuaded that the sound foundation behind the IOWA operator—and OWA operators in general—does serve as a solid platform to expand its use in consensus scenarios. The core of the additional research to be conducted would be on the role of α in the hardiness of the type of consensus being achieved.

References

1. Appel, O., Chiclana, F., Carter, J., & Fujita, H. (2016). A hybrid approach to sentiment analysis. In *IEEE: Proceedings of 2016 IEEE Congress on Evolutionary Computation (CEC): IEEE World Congress on Computational Intelligence (IEEE WCCI-2016)*, Vancouver, Canada, 24–29 July 2016, pp. 4950–4957.
2. Appel, O., Chiclana, F., Carter, J., & Fujita, H. (2016). A hybrid approach to the sentiment analysis problem at the sentence level. New avenues in knowledge bases for natural language processing. *Knowledge-Based Systems, 108*, 110–124.
3. Bishop, C. M. (2006). *Pattern recognition and machine learning*, 1st edn. LLC. Springer Science + Business Media.
4. Bordogna, G., & Sterlacchini, S. (2014). A multi criteria group decision making process based on the soft fusion of coherent evaluations of spatial alternatives. In L. A. Zadeh, A. M. Abbasov, R. R. Yager, S. N. Shahbazova, & M. Z. Reformat (Eds.), *Recent developments and new directions in soft computing* (Vol. 317, pp. 65–79). Studies in Fuzziness and Soft Computing. Springer International Publishing.
5. Boroushaki, S., & Malczewski, J. (2010). Using the fuzzy majority approach for GIS-based multicriteria group decision-making. *Computers & Geosciences, 36*(3), 302–312. Mar.
6. Chiclana, F., Herrera, F., & Herrera-Viedma, E. (1998). Integrating three representation models in fuzzy multipurpose decision making based on fuzzy preference relations. *Fuzzy Sets and Systems, 97*(1), 33–48.
7. Chiclana, F., Herrera-Viedma, E., Herrera, F., & Alonso, S. (2007). Some induced ordered weighted averaging operators and their use for solving group decision-making problems based on fuzzy preference relations. *European Journal of Operational Research, 182*(1), 383–399.
8. León, T., Ramón, N., Ruiz, J. L., & Sirvent, I. (2014). Using induced ordered weighted averaging (IOWA) operators for aggregation in cross-efficiency evaluations. *International Journal of Intelligent Systems, 29*(12), 1100–1116.
9. Pang, B., & Lee, L. (2008). *Opinion mining and sentiment analysis*. The essence of knowledge (Vol. 2, Nos. 1–2, pp. 1–135). Foundations and Trends in Information Retrieval. NOW
10. Pang, B., Lee, L., & Vaithyanathan, S. (2002). Thumbs up? Sentiment classification using machine learning techniques. In *Proceedings of the Association for Computational Linguistics (ACL-02) Conference on Empirical Methods in Natural Language Processing (EMNLP)* (Vol. 10, pp. 79–86).
11. Pasi, G., & Yager, R. R. (2006). Modeling the concept of majority opinion in group decision making. *Information Science, 176*(4), 390–414. Feb.
12. Pérez, L. G., Mata, F., & Chiclana, F. (2014). Social network decision making with linguistic trustworthiness-based induced OWA operators. *International Journal of Intelligent Systems, 29*(12), 1117–1137.
13. Perkins, J. (2010). *Python text processing with NLTK 2.0 cookbook*. Packt Publishing.
14. Yager, R. R. (1993). Families of OWA operators. *Fuzzy Sets and Systems, 59*(2), 125–148. Oct.
15. Yager, R. R., & Rybalov, A. (1996). Uninorm aggregation operators. *Fuzzy Sets and Systems, 80*(1), 111–120. Fuzzy Modeling.
16. Yager, R. R. (1988). On ordered weighted averaging aggregation operators in multicriteria decision making. *IEEE Transactions on Systems Man and Cybernetics, 18*(1), 183–190.
17. Yager, R. R. (2001). The power average operator. *Transactions on Systems, Man, and Cybernetics, Part A: Cybernetics, 31*, 724–730.
18. Yager, R. R., & Filev, D. P. (1999). Induced ordered weighted averaging operators. *IEEE Transactions on Systems, Man and Cybernetics, Part B: Cybernetics, 29*(2), 141–150.
19. Zadeh, L. A. (1983). A computational approach to fuzzy qualifiers in natural languages. *Computing and Mathematics with Applications, 9*(1), 149–184.
20. Zadeh, L. A. (1973). Outline of a new approach to the analysis of complex systems and decision processes. *IEEE Transactions on Systems, Man and Cybernetics, Part B-Cybernetics, SMC-3*, 28–44.

Fostering Positive Personalisation Through Fuzzy Clustering

Raymond Moodley

Abstract Elements of personalisation theory, personalisation engine modelling, and artificial intelligence algorithms using Fuzzy clustering are combined to provide a useful approach to enable positive personalisation that produces valuable outcomes for both the user and organisation. A recent case study in grocery retail, which applied this approach, shows that it is possible for a store to offer personalised promotions to its customers and in the process increase its market share and increase savings for customers. Adopting this personalisation approach creates a mutually beneficial environment, which is the essence of fostering positive personalisation.

Keywords Fuzzy C means · K-means · Clustering · Positive personalization · Recommender systems

1 Introduction

In December 2006, Time magazine awarded its coveted "Person of the Year" title to You, the people, for seizing the power of information from the few, and through the internet, making small contributions that collectively "changed the way in which the world changes". Time's technology editor at that time, Grossman [8], noted that at the heart of this, was Time's recognition of the power of individuals sharing their personal accounts, preferences, ideas of news, events, products, etcetera, and this, in its aggregated form, was becoming more powerful than any information source seen before. Fast forward fourteen years, and of course the people-driven information sector, which includes all forms of social media, is by far more influential than anything else that we know today. This sector has become so powerful that it has now given rise to a growing trend of a 'warped' or false reality, known as fake news. Fake news has not only changed the way in which people view events, products etcetera, but has changed their behaviour as well, and in some cases to their own detriment, for example, anti-vaccination campaigns, and climate change [14].

R. Moodley (✉)
Institute of Artificial Intelligence, De Montfort University, Leicester, UK
e-mail: raymond.moodley@my365.dmu.ac.uk

© Springer Nature Switzerland AG 2021
J. Carter et al. (eds.), *Fuzzy Logic*,
https://doi.org/10.1007/978-3-030-66474-9_4

This explosion of liberal mass sharing of data, in many cases very personal data, on a variety of information technology platforms, has resulted in organisations eyeing the opportunities that come with access to the thoughts and preferences of billions of people. As a result, data acquisition, ownership, and mining are now core operational components of most organisations, as it serves as a window into the lives of its stakeholders, from where it can expand its commercial and/or influential reach.

From the individual's perspective, there is an ongoing desire to achieve greater individualisation, that is, the promotion of the self as 'free to choose', self-directed, independent, and separate from others. One marker of societal evolution over the ages has been the loosening of the social structure, which was largely borne out of tradition. Thus, today, there is an ever-increasing choice with regards to the traditions that one must follow and adopt, and as a result, individuals are free to choose their own destiny, consequently increasing the state of individualisation within society [12].

Personalisation, in the context of modern society, may thus be seen as the intersection of the organisation's perspective and the individual's perspective. Done correctly, personalisation fosters a mutually beneficial relationship, or commonly referred to as a 'win–win'. This typically occurs when an organisation customises products or services to the needs of individual customers or groups of customers, using the data it has collected, and offers these products to these customers in a manner that is fair and reasonable. Positive personalisation is thus defined as the process of customising offerings, based on data, that results in a mutually beneficial outcome.

This chapter discusses key elements of personalisation and how this can be achieved using artificial intelligence (AI). Clustering, particularly fuzzy clustering, is discussed in detail, followed by a highlight of a real-life example in grocery retail that was extracted from Moodley et al. [10]. The chapter ends with a summary of key points.

2 Personalisation

Today, personalisation is ubiquitous thanks to the exponential rise of online transacting, and vast volumes of data sharing by consumers. For businesses, personalisation is no longer a luxury and has become a necessity for survival [6]. The benefits of personalisation have been realised from both consumers and businesses alike. From a consumer perspective, current statistics gathered from various research sources and collated in Forbes [6] show that over 90% of consumers prefer to engage with brands that offer them personalised content. Further, they are willing to share data if such experiences are easier and result in lower prices. While from a business perspective, Forbes [6] notes that as a result of personalisation, over 98% of businesses have seen an advance in customer relationships, with 95% of businesses recording increases in profitability after the first year of personalisation campaigns.

So, what is personalisation and why is it so effective? There is no precise definition of personalisation, but in general, there is consensus on the elements that define the concept. Oulasvirta and Blom [12] define personalisation as a "process that changes the functionality, interface, information content, or distinctiveness of a system" to an individual's preferences or needs. On a similar thread, Adomavicius and Tuzhilin [1] noted that personalisation involves the tailoring of products, services and information for an individual based on knowledge of their preferences (or needs) and behaviour within a given context. The notion of loyalty and trust also plays a vital role in successful personalisation. In this regard, Riecken [16] defined personalisation as a process of building loyalty through a meaningful one-to-one relationship. This is achieved by understanding the needs of an individual and then helping them fulfil these needs efficiently and knowledgably.

The effectiveness of personalisation stems from its roots in human psychology, and the notion that personalisation is a human need as opposed to a want. In this regard, Oulasvirta and Blom [12] used the theory of self-determination and its three subcategories (autonomy, competence and relatedness) to effectively demonstrate the underlying concepts that make personalisation so powerful, and indeed so popular.

Self-determination is the need to have one's choice, rather than having environmental elements determining one's actions. Thus, humans are motivated to take a certain action and persist in doing it if (1) it maintains an adequate level of stimulation, (2) it enhances competence and personal causation, or (3) if it propagates self-determination [12]. The three subcategories that underpin self-determination cement the link between self-determination theory and the need for personalisation. According to Oulasvirta and Blom [12], autonomy may be defined as the sense of unpressured willingness to engage in an activity, competence as the propensity of having an effect on the environment and to attain valued outcomes within it, whilst relatedness is the desire to have strong, close, warm bonds and relationships with other people (the desire for interconnectedness).

Based on the above, it is thus clear that humans have an innate need to personalise, and are likely to increasingly seek out, and gravitate towards products and services that will satisfy this need. It is thus only natural that organisations wishing to enhance their customer volumes and interactions become increasingly more personal in the way they design, produce and market their products and services.

2.1 How to Personalise?

The balance between personalisation, practicality and affordability is dynamic and indeed an ongoing challenge for organisations. Whilst industrialisation enhances affordability and practicality, it diminishes personalisation. Similarly, consumers prefer personalised products and services but willingly compromise as a result of affordability and practicality. Note that affordability is not the only driver or indeed the main driver among consumers, as other factors, e.g. time, quality, logistics and more recently data privacy are also important considerations in deciding the level of personalisation.

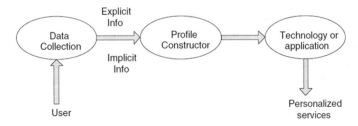

Fig. 1 User profile construction for personalisation, detailed in Bozdag [2]

The process of personalisation begins with the construction of a user profile. Figure 1 details the steps involved in constructing a user profile.

Irrespective of the approach used, the key first step is the data collection phase. Data can be explicitly provided, that is, the user either captures the data on to a system by themselves or tells the organisation specific, relevant details about themselves. For example, a curriculum vitae (CV) may be a piece of explicit data collection for a job search site, or a profile that a user creates for themselves on a dating or matchmaking site. On the other hand, implicit information gathering is usually very subtle and the user may not necessarily be aware that they are providing information, or that such information is being collected. This type of information gathering is controversial and was the basis of the General Data Protection Regulation (GDPR) adopted within the European Union in 2018. Implicit data is usually very powerful and, in many cases, includes data that is sub-consciously transmitted by the user. As a result, manipulation of such data to influence the user can also be very effective [2]. For example, if it is known that a person has been a victim of a traumatic house burglary, then both businesses and to some extent political parties can influence the behaviour of this person to their advantage without the user noticing it. For example, travel companies can emphasise low crime locations when this user searches for holidays, while political parties may target the user with campaign material that emphasises its strong stance on crime prevention during elections etcetera.

At this point, it is important to note that implicit data can be used both positively and negatively, and the intention of this chapter is to foster its positive use. Negative use continues, and as a result, users are becoming less trusting about sharing data, which benefits nobody in the long term. Data sharing is essential for innovation and progress and when this stops, it is likely that innovation will slow, and as a species, humans may progress slowly if not at all.

The profile constructor step, as shown in Fig. 1, is the heart of the user profiling process. In this step, the user's profile is built using any one, or a combination of machine learning algorithms. For example, a user profile database of a supermarket may implicitly conclude that a person is pregnant if the user is female, starts buying vitamins and unscented beauty products, and stops buying alcohol. This implicit profiling can be easily done using association rule mining (ARM).

The technology or application step entails the system that leverages the constructed profiles to deliver personalised services. This is usually some form of recommender system (RS) that queries the user profile database using a set of criteria.

Note that the impression of total personalisation can be achieved relatively easily using a combination of a relatively small portion of the products available, even in a very large user base. Consider a large supermarket in the UK that typically has over 10,000 individual items for sale in its stores. Using combination theory, it is thus possible to generate a unique voucher booklet of 4 vouchers for over 64.6 million users (the approximate population of the UK) using just 200 products. The supermarket can pick any 200 products to promote and the RS will query the user profile database to create these highly personalised vouchers that will be sent to each of the 64.6 million users it has on its database. For example, a pregnant user that likes chocolate may receive a voucher for vitamins, ham, chocolate and unscented lotion; whilst her pregnant friend that is vegetarian may have a voucher for cheese instead of ham. The above personalisation methodology is supervised learning, in that the database queries are fitted into predefined classes, with known labels [17]. This level of personalisation already occurs in supermarkets. American food retailer Kroger claims to have produced 11 million unique voucher booklets every 12 weeks for its customers, and the redemption rate from these campaigns is very high, with over 70% of customers using at least one voucher within 6 weeks [5].

However, it may be possible to personalise where the class labels are not predefined, for example, in the case of linguistic labels. Such personalisation leverages unsupervised learning methods, for example, clustering [17]. Consider the concept of loyalty—a supermarket may want to group its customers into three categories of loyalties based on their purchasing volume: low, medium, high, where the values associated with each of these labels are both unknown and dynamic. In this case, querying the database is not enough, and thus further processing is required on the data extracted from the user profile database. Clustering, including fuzzy clustering, is a powerful tool that enables personalisation using linguistic labels, and forms the basis of the case study that is discussed later in this chapter.

2.2 Models of Personalisation

Personalisation engines detailed in Adomavicius and Tuzhilin [1], is a "zoom-out" of the mechanisms detail in Fig. 1. The three models of personalisation detailed in Fig. 2 are highly relevant for the current context of continuous online transacting and data sharing. The model shown in Fig. 2a, c are the traditional models of personalisation that we all have become accustomed to, with the square between the providers and consumers being the personalisation engine. In Fig. 2a, the personalisation engine (provider centric) is typically owned by the provider, e.g. a grocery retailer, who collects data from its users, e.g. through a loyalty program, builds user profiles and uses these profiles as part of its marketing and store operation systems. This model is discussed as part of the case study detailed later in this chapter.

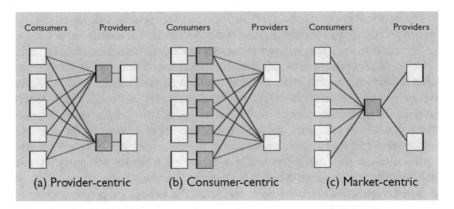

Fig. 2 Personalisation engines and models, detailed in Adomavicius and Tuzhilin [1]

In Fig. 2c, the personalisation (market centric) is typically developed by a third party, whose purpose may or may not be explicitly known. Marketing companies or agencies are examples of third parties, whose role is to help providers with market-based consumer preferences. These agencies conduct market research through surveys, consumer focus-group studies and consumer scanner panel analyses, and then use this insight to help providers produce, stock and sell products and services that appeal to the market. On the other hand, there are implicit personalisation engines that also play a role, and these are typically indirectly provided as a result of public sector organisations wanting to improve their operations. For example, a government census obtains demographic data of a local community which is used by government agencies to understand the demand for public services. However, providers can also leverage this data to understand the market in which they operate, and thus personalise their products and services to suit this market.

It is important to note that the way in which the provider-centric and market-centric models collect data limits contextual information being captured. Thus, these models do not provide a holistic view of the customer, including the motivations for their behaviour at that specific time. For example, consider a supermarket operating in a community where a large proportion of its residents do not eat beef for cultural reasons. Based on the market-centric engine, the store will typically stock very little beef products. A consumer buying flour in this store will be categorised as a consumer that likely uses flour for non-beef meals. The context as to why the consumer bought flour is not known.

The prolific rise in social media, and online search engine use has resulted in providers now being able to include context into personalisation. As a result, implicit personalisation is becoming increasingly possible, thus making marketing highly effective. This is achieved using consumer-centric personalisation engines, as shown in Fig. 2b. In this model, third-party data aggregators scrape a variety of data sources to build a profile for each consumer which is then sold to providers. Providers accessing these engines not only understand which product or service is required, but also the context behind this requirement. Using the earlier example of the consumer buying flour, a consumer-centric search engine may combine social media posts with search engine history to reveal that the consumer is planning on baking a birthday cake for their daughter's tenth birthday party. Armed with this knowledge, the provider can enhance the consumer's in-store experience by recommending and discounting related products, for example, age-appropriate cake decorations, soft drinks and other party paraphernalia. This approach may lead to a positive shopping experience for the consumer, an enhanced birthday party for their daughter, and thus may be seen as one example of positive personalisation. Note that this scenario could easily turn negative, where the provider, realising the intention of consumer, provides recommendations for higher priced party accompaniments, resulting in the shopper rejecting such recommendations, and either switching to another provider who is cheaper, and/ or losing trust in the provider altogether. Clearly, such practices are a "lose-lose" scenario, and fosters an environment of mistrust of personalisation technology, sometimes referred to as "creep factor".

3 Clustering

The process of grouping data objects together in some meaningful way, be it through clustering or classification, is considered to be a knowledge discovery process, in that one could draw new conclusions about the data set following this process, that was not previously possible [17]. In essence, this is the mechanism that enables personalisation, as discussed in the previous sections.

Cluster analysis, or clustering, may be concisely defined as the process of grouping physical or abstract objects into classes, such that objects within a class or cluster are similar to each other (intra-cluster similarity) but dissimilar to objects in other clusters (inter-cluster dissimilarity). The values or labels for each of these classes are not fixed, but dependent on the context and the data set. Note that this is the key difference between clustering and classification. In classification, objects are also grouped in classes, however the class labels are known, and thus the process may be considered to be a data "fitting" exercise [9, 17].

Given that clustering does not have fixed class labels, it thus becomes a powerful tool in grouping categorical data, including linguistic data. This type of grouping is typically done by humans on a daily basis to perform even some of the most mundane of tasks that computers will find difficult, if not impossible, to do. The following two examples are used to illustrate these concepts.

Example 1 Consider the medical task of grouping people based on their body temperature. A medical practitioner may group people with a body temperature of above 37.8 °C in a class labelled "fever", those with a temperature below 35 °C as "hypothermia", and finally those with a temperature in between 35 °C and 37.8 °C as "normal". These are well-defined classes, with well-defined labels and thus represents a classification problem or a supervised learning task.

Example 2 Suppose a sample of people were asked to label a series of outside temperatures as either "cold", or "pleasant", or "hot", and based on these responses, we are required to determine the temperature for labels for each of these classes. Clearly, there are several other factors that will influence these labels, including the location that is associated with these temperatures and the home-location of the person providing a response. For example, a temperature of 18 °C in Glasgow may be classed as "pleasant" by a Glaswegian but will likely be classified as "cold" by a Dubaian. Given that the class labels are not fixed, the solution to this problem is usually context dependent, and is better solved using clustering (unsupervised learning).

3.1 Crisp Clustering Versus Fuzzy Clustering

The ideal grouping of objects in a two-dimensional plane consisting of three clusters may be represented by an equilateral triangle, with each apex being a cluster centroid (or sometimes called a cluster prototype), and all data points tightly packed around any of these centroids. However, in real-life, the situation is quite different, and very often data points will straddle the boundary that exists between clusters. Irrespective of the clustering algorithm used, it is sometimes possible to have two or more data points, each belonging to a different cluster, lying close to the cluster boundary. In fact, these data points can sometimes be closer to each other than to their respective centroids. In the case of crisp clustering, which obeys binary logic and conventional set theory, the data points will be assigned to their respective clusters and will have no connection with the other clusters. However, in fuzzy clustering, those laws of fuzzy logic are applied, and each data point is assigned a membership weight to each cluster, thus theoretically belonging to every cluster to a varying degree. The data point is then ultimately assigned to the cluster with the largest membership weight, that is "the most possible" cluster. Note that the final answer in crisp and fuzzy clustering may be the same in that the data points will be assigned to the same respective clusters, but it is the approach that is fundamentally different. Fuzzy clustering attempts to mimic a more human-like approach, where the awareness of alternatives, and the intensity of the choice-making is considered, thus allowing for greater control.

Consider the example of rounding the numbers 14.9 and 15.1 to the nearest multiple of ten. Given this problem to solve, a ten-year-old child may immediately use the mathematics that they have been taught and round down 14.9–10 and round-up 15.1–20. There may be some human awareness that 14.9 and 15.1 are very close to each other, but it is likely that the child will ignore this fact and answer the question that was posed. In this instance, the child's processing may be considered as being crisp. We also note that if 10 and 20 are the centroids of two clusters, then 14.9 and 15.1 are likely to be data points that sit on the cluster boundary and significantly closer to each other than they are to their respective centroids.

Now consider a different scenario involving the two numbers where students have taken an important test where the pass mark is 15. The student obtaining 15.1 will be happy that they have passed, but acutely aware, and most likely relieved, that they only just passed. On the other hand, the student that obtained 14.9 will be sad that they have failed, but again acutely aware as to how close they were to passing. Thus, we note that the outcome itself is crisp, that is pass or fail without any awareness of the closeness of the mark, but the human element is fuzzy, as in both cases the student was fully aware of the closeness of their respective mark. This human element may also be picked-up in other human-related judgements with regards to these marks. For example, the lecturer may ask the student that passed to attend additional tutorials to firm-up their knowledge whilst proceeding with the next module. Similarly, the lecturer may allow the student who failed to proceed with the next module while attend additional tutorials to firm-up their knowledge. As noted in this example, human intervention (the lecturer's decision) has resulted in similar outcomes for both the 14.9 and 15.1 cases.

3.2 The K-Means and Fuzzy C-Means (FCM) Clustering

Several clustering algorithms exist, each with its pros and cons, and are well-documents including in Han et al. [9], and Tan et al. [17]. However, to illustrate the differences between traditional, crisp clustering, and fuzzy, or soft clustering, we will compare the basic algorithms for two popular clustering approaches, K-Means (crisp clustering) and FCM (fuzzy clustering).

These two algorithms have been the subject of comparison in several studies notably in Cebeci et al. [3], Panda et al. [13], and Ghosh et al. [7]. Whilst it is well-established that K-Means is superior in terms of speed and requires less processing power, the choice between K-Means and FCM remains application and context specific. Indeed, in some applications, where the clusters were well-separated and the data sets were large, K-Means showed superior performance, whilst FCM showed better performance in datasets that were noisy, as is typically found in human inter-action data. In any event, studies have generally concluded that for most data sets, the use of K-Means should be a good starting point, and depending on the results, FCM or other techniques should be considered.

3.3 The K-Means and Fuzzy C-Means (FCM) Algorithms

Suppose we have n data points in a data set, $D = \{x_1, x_2, x_3, \cdots, x_m\}$ and we wish to cluster these points into k clusters, such that n_i is the number of data points assigned to the ith cluster, c_i. The K-Means clustering algorithm is detailed in Algorithm 1, whilst Algorithm 2 details the Fuzzy C-Means clustering.

Algorithm 1: K-Means Clustering

1: Select k points to represent the initial cluster centroids.
2: Assign each data point to its nearest centroid using the Euclidean distance measure, $d(x_i, c_i) = x$.
3: Update the centroids to be the mean distance of all data points assigned to it, using $c_k = \frac{1}{n_k} \sum_{x \in c_i} x$.
4: Repeat steps 2 and 3 as some data points may be closer to another centroid, and consequently, re-assigned.
5: Stop when no further re-assignments are made, that is, the sum of squared errors (SSE) of all data points is minimised, $SSE = \sum_{i=1}^{k} \sum_{x \in c_i} x^2$.

Algorithm 2: Fuzzy C-Means Clustering (FCM)

1: Randomly initialise a membership matrix, U^0, comprising of n rows and k columns, with each matrix element, u_{ij}, representing the fractional membership of data point, i, to cluster, c_j.
2: Determine the cluster centroid, c_j, for each cluster using the product of Euclidean distance of each data point to the centroid and the weighted-average fractional membership of that cluster, that is, $c_j = \frac{\sum_{i=1}^{n} u_{ij}^m x_j}{\sum_{i=1}^{n} u_{ij}^m}$. Note that m is a fuzzifier that denotes the "vagueness" in the data, such that $m \in [1, \infty)$; generally $m = 2$.
3: Update the membership matrix, U^t, where element $u_{ij} = \dfrac{1}{\sum_{p=1}^{k} (\frac{x_{ij}}{x_{ip}})^{\frac{2}{(m-1)}}}$ and $1 \leq t \leq z$, where z is the total number of iteration.
4: Repeat steps 2 and 3 unit $|U^{(t+1)} - U^t| \leq \varepsilon$, where ε is a predefined convergence value. Note that if convergence is not reached before t reaches z, then it may be necessary to adjust z or ε.
5: An alternative approach, which will yield similar results, is to minimise an objective function, F, which is based on the sum of squared errors (SSE) of all data points and the membership function, that is: $F = \sum_{i=1}^{n} \sum_{j=1}^{k} u_{ij}{}^m x_{ij}$, hence when F is minimised, i.e. $F^{(t+1)} = F^{(t)}$ for some large value of t.

The above two algorithms have been used to develop packages for programming software including Python and R, thus making the implementation of clustering into larger data processing applications relatively straightforward.

4 Case Study: FCM Clustering to Enable Personalisation for Targeted Promotions in Grocery Retail

The use of FCM in a real-life application is demonstrated leveraging the work detailed in Moodley et al. [10]. The grocery retail business is vast, highly competitive and indeed can sometimes be quite fickle. Gaining and retaining market share, which is typically an indicator of business performance, often hinges on attracting new customers, and retaining existing ones, whilst minimising the associated costs. One popular way of achieving this goal is by offering promotions to a select group of customers (targeted promotions), who would typically be less inclined to buy the product at the given retailer, or at all for that matter. Promoting in this manner ensures that customers who regularly purchase the product are not unnecessarily given a discount, thus preventing generalised price reductions, which can be detrimental to the retailer, especially in a sector that operates on very tight profit margins, typically less than 3% in the UK.

At this point, one can argue that this practice violates the positive personalization premise that this chapter is reinforcing, as loyalty is being punished. No doubt we have all experienced at some points in our lives, companies offering attractive promotions to new customers that are not offered to loyal customers, which in this case is usually us. It is important to realise that there are many factors of loyalty beyond price, and whilst some of these are intangible, at some point human nature overrides these factors, and one may switch loyalties to a product or service, either temporarily or permanently. Some of these factors are well-described, in the context of grocery retail, in Rhee and Bell [15]. Hence, for example, a customer may be loyal to a retailer because it is convenient, or if it has an appealing store layout, or if it has ample parking. Even though the customer may know that prices are higher at this retailer, they will forego this in order to capitalise on the other factors that they consider valuable. At the same time, if this customer needed a single product, for example, milk, and was far away from their usual retailer, then they may buy this product at a retailer close to them on this occasion, and then revert to their usual buying behaviour at some point in the future. Thus, the concept of loyalty is dynamic, with people regularly evaluating trade-offs between factors in order to maximise their own utility. Indeed, price is usually the largest factor of loyalty, but in many cases, not the only factor. However, being generally the largest factor, it has also become the most popular factor in terms of targeted promotions.

FCM, which is part of the application/ technology step in Fig. 1, is used to stratify customers into several bands, based on their purchasing history of a given product. However, designing a stratification strategy based on a customer's purchasing history can be complex, as all customers are typically different. To reduce complexity, the well known RFM (Recency, Frequency and Monetary) framework for customer targeting in a retail setting was used as the starting point [4].

4.1 Identifying Target Customers

Every customer of a retailer can be placed along the RFM continuum for every item within the retailer's store. The RFM continuum can be visualised as a three-dimensional plane with the three axes representing R, F and M. Clearly this concept can be become very complex given that large retailers typically have millions of individual customers and sell thousands of products. As a result, this concept has been simplified to enable faster processing, whilst enabling a good degree of stratification.

"Recency" is considered an elimination variable, with the assumption for this scenario being that customers must have made at least two purchases during the period under consideration, with at least one containing the target item, else they are not selected. This assumption can be adapted to the sector under consideration, and in the retail sector, making at least one purchase of the target item is appropriate to reduce "false positives". There may be some customers who may be appropriate targets, and who have not purchased the target item in the past, but this introduces several variables that are difficult to measure, including taste, allergy/intolerance, and cultural aversion, and may result in increased "false positives", which can be detrimental to the retailer.

"Frequency" is item-specific and is based on the number of times that a customer has purchased the target item. In this regard, the concept of support and confidence, as detailed in ARM theory has been used as it is a simple, yet effective way of introducing quantification of categorical variables. As noted in Moodley et al. [10], the support of the target item, C, denoted by supp(C), such that $0 \leq supp(C) \leq 1$, is the number of times a customer purchases C in a given period, T, where $T = \{T_1, T_2, T_3, ..., T_N\}$, is the set of N equal and distinct sub-periods. In practical terms, T is typically one calendar year, and the sub-period, T_i, is a week. Hence, all shopping by a customer during a week is typically considered as one transaction. Further, the confidence of the purchase of item A leading to the purchase of item C, conf(A → C), is defined as supp(C)/supp (A and C).

"Monetary" is defined as the relative average size of the customer's transactions. The notion of relative average size takes into consideration the household size that the customer represents. Note that the intention of customer stratification is to group customers in bands of loyalty, with the least-loyal customer bands becoming priority targets as they have the potential to create new revenue for the retailer. Whilst larger households typically spend more in absolute terms, these households may not necessarily be loyal in relative terms as it may be perfectly plausible that a single or two-person household could spend less in absolute terms but purchase all of their groceries from the given retailer. As a result, the "Monetary" dimension for customer U, is better defined as the average customer transaction size for retailer, T_{Uav}, divided by the household size, f_U, for customer, U.

4.2 Creating a Target Customer Stratification Matrix

Given that "Recency" in this scenario was considered an elimination variable, with binary attributes, with a "Recency" = 0 being eliminated, the customer stratification matrix is thus a two-dimensional matrix, comprising of the "Frequency" and "Monetary" dimensions for just those customers with "Recency" = 1. "Frequency" and "Monetary" can each be divided into several mutually exclusive segments, noting that the larger the number of divisions, the larger the amount of processing required, and the better the personalisation. In this example, "Frequency" and "Monetary" are each divided into three categories, "Low", "Medium", and "High" as shown in Table 1.

Table 1 Monetary/Frequency mapping of customers

	"Frequency"		
"Monetary"	Low, Low	Low, Medium	Low, High
	Medium, Low	Medium, Medium	Medium, High
	High, Low	High, Medium	High, High

4.3 Customer Treatment Approach

The treatment approach adopted by the provider is entirely up to that provider. In this example, the mapping in Table 1 was used to create nine clusters, as shown in Table 2. Customers were considered to be "Switchers" if their frequency was less than their monetary spend, except for cluster one where both frequency and monetary spend were "Low". These customers typically conduct a large proportion of their purchases at other stores and/or have a low take-up of the targeted item in this store. Customers were considered "Loyal" when their frequency and monetary spend were similar and at least "Medium". These customers conduct most of their shopping at this store. Customers that had a "High" or "Medium" frequency with a low monetary spend were potential "Drop-Outs" as they appeared to spend elsewhere but have a high take-up of the target item within this store. As a result, these customers could be enticed to take their custom for the target product to another store. It should be noted that customers in Cluster 6 were considered to be "Switcher" because their frequency was less than their monetary spend, implying that these customers chose to purchase the target item elsewhere, even though they were essentially loyal to the store for other purchases. Consequently, the right incentive may enable customers in Cluster 6 to switch purchases away from other stores to this store.

Table 2 Customer treatment approach

Cluster	Frequency	Monetary	Approach
1	Low	Low	Switcher
2	Low	Medium	Switcher
3	Low	High	Switcher
4	Medium	Low	Drop-out
5	Medium	Medium	Loyal
6	Medium	High	Switcher
7	High	Low	Drop-outs
8	High	Medium	Loyal
9	High	High	Loyal

4.4   Creating and Using Clusters for Personalisation

Nine clusters were created using a two-step fuzzy clustering process as shown in Algorithm 2 and based on the approach detailed in Table 2.

Algorithm 3: Creating Nine Treatments Clusters using FCM

For each item, i, in Store, S,

1: Cluster the Monetary parameter, denoted by $\frac{T_{Ui(avg)}}{f_{Ui}}$ where $T_{Ui(avg)}$ is the average transaction size

of the consumer, U_i, in this store and f_U is the consumer's household size, into three clusters ("Low", "Medium", "High") using Algorithm 2. Note that this normalisation is important to cater for differing household sizes.

2: The objective function to be minimised in this instance is $F = \sum_{i=1}^{n} \sum_{j=1}^{3} u_{ij}{}^m d\left(\frac{T_{Ui(avg)}}{f_{Ui}} - c_j\right)$,

where c_j is the centroid of cluster, j, and n is the total number of users.

3: For each of the clusters created in Step 2, cluster further into three clusters ("Low", "Medium", "High") based on $supp(C)$, the frequency of purchasing the target item, A. The concept of support is detailed in Moodley et al. (2019).

4: The objective function to be minimised in this instance is $E = \sum_{i=1}^{p} \sum_{j=1}^{3} u_{ij}{}^m d\left(supp(C) - c_j\right)$,

where c_j is the centroid of cluster, j, and p is the cardinality of each $\frac{T_{Ui(avg)}}{f_{Ui}}$ cluster.

The resulting nine clusters provide the customer groupings for all customers that have been considered for targeted promotions for the target item. This can be repeated for all products that the store wants to promote. For example, customer number 102 who has "medium" monetary spend in the store, may be grouped as "Loyal" for items C and D, whilst be grouped as "Switcher" for items F and G. Thus, it is now possible for the store to create a 4-leaf voucher booklet (one leaf for each item) for customer 102, with differing promotion options for each item. The items where customer 102 has been classed as "Loyal" could have a lower discount than the items where the customer is classed as "Switcher". As a result, and as discussed in Sect. 2.2, the voucher booklet issued to customer 102 is highly personalised as it may be very different from booklets offered to other customers. Notice the "win–win" nature of this personalisation which fosters a positive environment. First, the store has managed to identify opportunities to upsell products to its existing customers by offering larger discounts on items it believes that they buy elsewhere, whilst sustaining loyalty on products that they already buy in their store. This will manifest in a revenue upside for the store. Second, the consumer feels rewarded through receiving the voucher booklet, and is further delighted that the booklet contains vouchers for products that they actually use compared with items they never use or have never even heard of. This is likely to result in the customer increasing their purchasing at the store, and sharing their data, e.g. through a loyalty program, to enable them to receive further vouchers for other products that they may be using.

4.5 Measuring the Accuracy of the Personalisation Approach

Formal methods for calculating clustering efficiency are available in data mining texts, including in Han et al. [9] and Tan et al. [17]. However, these methods will only evaluate the effectiveness of the clustering and not whether the overall business objective has been met (either partially or fully). To achieve this objective, the accuracy of the personalisation approach may be easily verified using either internal and/or external data. From an internal perspective, the change in loyalty of targeted customers can be monitored over time. For example, customer 102, who was identified as a "Switcher" in period 1 for product G can be re-evaluated in period 12. If purchasing has shifted towards "Loyal", then one can conclude that the personalisation approach is effective. Indeed, this can be turned into an aggregate metric across all customers, and thus could serve as a performance indicator for the store's personalised promotions program. This metric could be of the form: percent shift in "Loyalty" across all products and customers and is given by the total number of shifts in a 12-week period/total of number of vouchers. Thus, the store's performance is enhanced as this metric approaches 1.

The store could also adopt an external perspective, in which case it could engage third-party providers of consumer scanner panels. This approach is detailed in Moodley [11], where scanner panel data is used to test the effectiveness of the personalisation approach using the percent difference in mean purchasing (internal versus external) for each loyalty group. The advantage of this method is that it could be used during the design phase of the personalisation approach, thereby allowing for early adjustments. However, it could be more costly than the internal perspective due to costs associated with engaging the third-party provider. The results of applying this method on three anonymised products, as noted in Moodley [11] are shown in Fig. 3.

From Fig. 3, it can be clearly seen that the mean number of purchases made within a store (internal) is significantly higher for groups that are loyal as compared with those that are classified as "Switchers". Given this, it can be concluded that the clustering approach adequately segregates customers based on loyalty, thus providing the basis from which personalisation can take place.

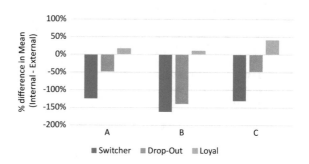

Fig. 3 Percent difference in mean purchasing by loyalty groups on three anonymised products. Adapted from Moodley [11]

5   Summary of Key Points

The summary of key concepts discussed in this chapter are as follows:

- Personalisation may be classified as a human need and is now a key ingredient for running a successful commercial or public sector organisation.
- Positive personalisation is the use of personalisation to create "win–win" value propositions for customers and organisations.
- Developing a robust personalisation engine includes collecting (implicitly and explicitly), storing, and processing user data to provide meaningful, customised, and positive interactions with users, products/services, and organisations.
- Artificial Intelligence, notably fuzzy clustering, is a useful tool in enabling and simplifying personalisation through the formation of groups that contain associated products/services, and users with similar preferences.

References

1. Adomavicius, G., & Tuzhilin, A. (2005). Personalization technologies: A process-oriented perspective. *Communications of the ACM, 48*(10), 83–90.
2. Bozdag, E. (2013). Bias in algorithmic filtering and personalization. *Ethics and Information Technology, 15*(3), 209–227.
3. Cebeci, Z., & Yildiz, F. (2015). Comparison of k-means and fuzzy c-means algorithms on different cluster structures. *Agrárinformatika/Journal of Agricultural Informatics, 6*(3), 13–23.
4. Fader, P. S., Hardie, B. G., & Lee, K. L. (2005). RFM and CLV: Using iso-value curves for customer base analysis. *Journal of Marketing Research, 42*(4), 415–430.
5. Forbes. (2013). Kroger knows your shopping patterns better than you do. https://www.forbes.com/sites/tomgroenfeldt/2013/10/28/kroger-knows-your-shopping-patterns-better-than-you-do/#156d3168746a.
6. Forbes. (2020). Stats showing the power of personalization. https://www.forbes.com/sites/blakemorgan/2020/02/18/50-stats-showing-the-power-of-personalization/#5ecd84c72a94.
7. Ghosh, S., & Dubey, S. K. (2013). Comparative analysis of k-means and fuzzy c-means algorithms. *International Journal of Advanced Computer Science and Applications, 4*(4).
8. Grossman, L. (2006). Time magazine person of the year. https://www.nbcnews.com/id/16242528/ns/us_news-life/t/time-magazines-person-year-you/#.Xq_W4f9KiUk.
9. Han, J., Kamber, M., & Pei, J. (2012). *Data mining: Concepts and techniques* (Vol. 10). Waltham, MA: Morgan Kaufman Publishers. 978-1.
10. Moodley, R., Chiclana, F., Caraffini, F., & Carter, J. (2019). A product-centric data mining algorithm for targeted promotions. *Journal of Retailing and Consumer Services*, 101940.
11. Moodley, R. (2019). Enhancing the prediction of missing targeted items from the transactions of frequent, known users.
12. Oulasvirta, A., & Blom, J. (2008). Motivations in personalisation behaviour. *Interacting with Computers, 20*(1), 1–16.
13. Panda, S., Sahu, S., Jena, P., & Chattopadhyay, S. (2012). Comparing fuzzy-C means and K-means clustering techniques: A comprehensive study. In *Advances in computer science, engineering & applications* (pp. 451–460). Berlin, Heidelberg: Springer.
14. Pulido, C. M., Ruiz-Eugenio, L., Redondo-Sama, G., & Villarejo-Carballido, B. (2020). A new application of social impact in social media for overcoming fake news in health. *International Journal of Environmental Research and Public Health, 17*(7), 2430.

15. Rhee, H., & Bell, D. R. (2002). The inter-store mobility of supermarket shoppers. *Journal of Retailing, 78*(4), 225–237.
16. Riecken, D. (2000). Introduction: Personalized views of personalization. *Communications of the ACM, 43*(8), 26–28.
17. Tan, P. N., Steinbach, M., & Kumar, V. (2016). *Introduction to data mining.* Pearson Education India.

Diagnosing Alzheimer's Disease Using a Self-organising Fuzzy Classifier

Jonathan Stirling, Tianhua Chen, and Magda Bucholc

Abstract Dementia is one of the major causes of disability and dependency among older people worldwide. Without treatment currently available to cure dementia or to alter its progressive course, one of the principal goals for dementia care set by the World Health Organization is the early diagnosis in order to promote early and optimal management. In recognition of the potentials of fuzzy systems in effectively dealing with medical data, this chapter investigates the use of a very recently proposed Self-Organising Fuzzy (SOF) classifier for the prediction of Alzheimer's Disease against Mild Cognitive Impairment and being Cognitively Unimpaired with patient observations provided by the renowned Alzheimer's Disease Neuroimaging Initiative repository. The experimental study demonstrates the effectiveness of SOF, especially in combined use with the Recursive Feature Elimination feature selection.

Keywords Dementia diagnosis · Alzheimer's disease · Clinical decision support · Fuzzy systems · Self-organising fuzzy classifier · SOF

For the Alzheimer's Disease Neuroimaging Initiative. (Data used in preparation of this article were obtained from the Alzheimer's Disease Neuroimaging Initiative (ADNI) database (adni.loni.usc.edu). As such, the investigators within the ADNI contributed to the design and implementation of ADNI and/or provided data but did not participate in analysis or writing of this report. A complete listing of ADNI investigators can be found at: http://adni.loni.usc.edu/wp-content/uploads/how_to_apply/ADNI_Acknowledgement_List.pdf.)

J. Stirling · T. Chen (✉)
Department of Computer Science, School of Computing and Engineering, University of Huddersfield, Huddersfield, UK
e-mail: T.Chen@hud.ac.uk

M. Bucholc
Intelligent Systems Research Centre, School of Computing, Engineering & Intelligent Systems, Ulster University, Londonderry, UK
e-mail: m.bucholc@ulster.ac.uk

© Springer Nature Switzerland AG 2021
J. Carter et al. (eds.), *Fuzzy Logic*,
https://doi.org/10.1007/978-3-030-66474-9_5

69

1   Introduction

Dementia is a progressive condition with an estimated 50 million cases worldwide in 2018 which is expected to more than triple to 152 million by 2050 [1]. Currently, between 60 and 70% of dementia cases are attributed to Alzheimer's Disease (AD) [2] with the remainder consisting of different types including Vascular and Frontotemporal Dementia, each with different causes. Medical research into dementia continues with new types such as Limbic-predominant Age-related TDP-43 Encephalopathy (LATE) still being discovered [3].

Dementia can be generally defined as a condition that impairs the regular cognitive functions of the brain [2]. This impairment affects individuals differently and with varying severity, but typically affects memory, language, behaviour and the ability to carry out day-to-day tasks [2, 4]. Dementia primarily affects older people but can also affect younger people with an estimated 40 thousand cases in the UK alone for age groups under 65 [5]. With increases in life expectancy resulting in larger aged populations, the effect of dementia is expected to have significant implications for economies, healthcare services and society in addition to the substantial physical, psychological and social impact it has on sufferers, their families, friends and carers [2].

Given the impact of and lack of a cure for dementia, it is important to diagnose those affected as early as possible and additionally target those at the highest risk. Diagnosis allows symptom-slowing medication regimes to be used [6] and for patients in combination with their local healthcare services to prepare care plans to preserve as high a quality of life as possible, for as long as possible [7]. Presently, diagnosis tends to occur late due to: manual diagnosis being time-consuming [8]; a lack of practitioners' confidence and/or training to be able to make correct decisions [9]; the limited amount of time in primary care patient interactions [10]; and waiting times up to 18 weeks in the UK [11]. Diagnosis is further complicated by diseases which can show similar symptoms to dementia as well as natural degeneration due to old age [4].

Recent research covers a range of data modalities and machine learning (ML) techniques when considering the diagnosis and pathology of dementia particularly with respect to AD. Of particular relevance for this work are those that consider AD prediction for individual patients based upon data including neuroimaging, neurocognitive assessments and other biomarkers. In general, the use of multiple assessments have been shown to be a good indicator of AD [8, 12], while considering assessments individually do not do as well (e.g., when considering only the Mini-Mental State Examination (MMSE) [13], it only lead to less than 0.7 accuracy using SVM and MLP.)

Additionally, there is no global standard for what tests are applied to patients between regions as shown by differences in those available via ADNI and those described in literature [8, 14]. For instance, [8] considers the creation of an aggregated questionnaire, built from selected questions from multiple tests, applying multiple ML techniques as well as FS with the intention to identify a single optimal questionnaire for dementia diagnosis. The simultaneous use of multiple assessments has also been conducted in [12], with SVM models producing their best results on a feature subset consisting of four clinical assessments providing multi-class accuracy of 83% and AUC of 95%, though their work also considers results from other modalities including MRI/PET and CSF biomarker data.

Among recent advances in machine learning that have had success in healthcare domains [15], fuzzy systems, which are built on top of fuzzy sets that permit gradual assessment of set elements, enable the tolerance of uncertainty and imprecision that may result from linguistic descriptions while enquiring medical symptoms or noise that may result from inaccurate testing results. While been widely applied in various domains [16, 17], fuzzy techniques have also been intensively utilised in numerous medical applications to tackle challenges raising from healthcare, (e.g. [18–20]). However, the application of fuzzy systems in diagnosing dementia is relatively limited in the literature. For instance, a recent work has attempted the Fuzzy Logic and Adaptive Neuro-Fuzzy inference Systems [7], however, the dataset used was severely limited in numbers of features and observations leaving questions as to whether the results were flawed.

As such, in working towards providing assistance for clinicians to conduct an effective diagnosis of Alzheimer's Disease, this chapter, therefore, proposes to utilise a very recently proposed Self-Organising Fuzzy (SOF) classifier [21] for the prediction of Alzheimer's Disease (AD) against Mild Cognitive Impairment (MCI) and being Cognitively Unimpaired (CU) patient observations. The patients' data used in this research comes from the renowned Alzheimer's Disease Neuroimaging Initiative (ADNI) repository. The underlying testing bed is a group of 488 patients and 66 variables, with feature selection methods also applied to explore the effectiveness of selected variables in the experimental study.

The remainder of this chapter is organised as follows. Section 2 introduces the background of the SOF and the summary of the data set used. Section 3 describes the proposed pipeline. Section 4 presents and discusses the experimental outcomes. Section 5 concludes the chapter and outlines ideas for further development.

2 Preliminary

2.1 The Self-organising Fuzzy Classifier

The Self-Organising Fuzzy (SOF) [21] is a non-parametric machine learning approach that considers dual-phase training made up of offline (initial) and online (run-time) phases and works based upon the creation of computed centres of data clouds (prototypes) and distance measures. This work considers the Euclidean distances of points in the offline phase of the technique for initial model development and testing due to the limited amount of data, and time constraints to implement data-streaming services.

The offline training of SOF aims to generate zero-order AnYa type fuzzy rules for every unique class in the dataset which take the form of a series of disjunctions of similarities, or fuzzy membership degree between an input vector and "prototypes" of each class. A zero-order AnYa type fuzzy rule has the following form:

$$IF \ (x \ \sim \ p_1^c) \ OR \ ... \ (x \ \sim \ p_n^c)$$
$$OR \ ... \ (x \ \sim \ p_N^c) \ THEN \ (class) \tag{1}$$

where x is the input vector; $\sim$ denotes similarity, which can also be seen as a fuzzy degree of satisfaction; p_n^c represents the n-th prototype for class c;

While other techniques build models for all classes combined, SOF training applies to subsets of the dataset split by the class of each observation, training each set independently with no interference between them. To create these fuzzy rule sets prototypes are derived from the unique samples for each class. For each sample, the multi-modal density is computed and the sample with the maximum density is added to a new list. The remainder of the samples is added to the list recursively, selecting the one with the minimum distance to the sample at the tail end of the list, noting that points cannot appear multiple times on the list.

From each class list, samples with a higher density than those immediately before and after them in the list are added as initial centroids. The items in each class list that were not selected are then used to form into data clouds around those centroids, with each sample belonging to the one closest to them. The centre of each cloud is determined and the density is computed using the number of samples in each cloud as a weighting. Neighbours of each cloud centroid are then computed based on whether the square distance between two of them is within a computed threshold based upon a user-provided granularity and the distances between points, after which the centroids with the highest density within a class neighbourhood are chosen as the prototypes for use in generating the AnYa rules. With respect to the granularity of the technique, the higher the value provided, the more prototypes are expected to be created resulting in a more finely defined area class area.

With respect to classification, once the model has been trained class predictions are made in two parts. First, a local decision is made for each class, resulting in an output of the strength of the data point per class by taking the negative square distance between the new observation x and each prototype as exponent to the Euler's constant. The second part selects the maximum strength calculated from all rules to determine the final classification, effectively choosing the single closest prototype to the sample. As such, this work considers whether the use of SOF can predict whether a patient observation can be correctly diagnosed as CU, MCI or AD when considering a subset of ADNI data.

2.2   Data Summary

Data used in the preparation of this article were obtained from the Alzheimer's Disease Neuroimaging Initiative (ADNI) database (adni.loni.usc.edu). The ADNI was launched in 2003 as a public-private partnership, led by Principal Investigator Michael W. Weiner, MD. The primary goal of ADNI has been to test whether serial magnetic resonance imaging (MRI), positron emission tomography (PET), other biological markers, and clinical and neuropsychological assessment can be combined to measure the progression of mild cognitive impairment (MCI)) and early Alzheimer's disease (AD). For up-to-date information, see www.adni-info.org.

In this research, aside from the decision variable, the selection of specific predictors from the ADNI repository follows that as described in the recent work by [12], which is briefly summarised as follows.

The dataset used comprises of baseline patient records with any incomplete observations (across all selected variables) having been removed from the set. The resulting dataset comprises of 488 observations split between AD ($n = 76$), MCI ($n = 218$) and CU ($n = 194$). Each observation in the dataset consists of 66 independent variables made up of 26 discrete and 40 continuous items and two dependent variables, one continuous and one discrete. The variables are split by the mode of testing used to extract the values. First, a patient data modality consists of demographic information including age, gender, education level, marital status, ethnicity and race. It also covers 19 items from the patient's medical history (including alcohol/drug abuse, smoking any cardiovascular or psychological issues) and expands to family history, describing if either parent or any sibling suffers from dementia.

A Clinical Measures (CM) modality covers the results of various test suites designed to determine issues with cognitive functions such as memory, learning and language. Specifically, this dataset collates the results from Alzheimer's Disease Assessment Scale 13 (ADAS-13), MMSE, three RAVLT trials, Functional Assessment Questionnaire (FAQ), Montreal Cognitive Assessment (MoCA) and two logical memory tests for immediate (LIMM) and delayed recall (LDEL), each of which tests specific areas of brain function.

MRI and PET scans make up a neuroimaging modality covering 23 of the variables provided in the dataset. The MRI measures describe the volumes of seven areas of the brain along with the volumes of white and grey matter, white matter hyperintensities, cerebrospinal fluid and intercranial volume. MRI data is also used to extract boundary shift integral values for the whole brain and ventricles. PET scans, depending on the tracer compound used, measure: glucose metabolism across five regions of the brain and overall five; the mean uptake of tracer; and the sums of pixels Z-scores two or three standard deviations from 0. MRI and PET data can be considered as independent modalities within neuroimaging.

Finally, CSF extraction as an independent and invasive procedure is considered an independent modality. The measures extracted from the fluid consist of the concentrations of certain proteins and the ratios between them.

Fig. 1 ML pipeline

With respect to dependent variables, the clinical decision for baseline observations is provided from ADNI, consisting of three classes describing CU, MCI and AD. It is worth noting that [12] made use of the Clinical Dementia Rating Sum of Boxes (CDRSB) continuous value as the dependent for regression and created their own classification variable based on CDRSB using two thresholds for the classification task. Whereas this work only looks at classification rather than regression problem, and to keep the feature parity with the original work, the CDRSB variable is ignored as an independent feature, while the categorical diagnosis alone is used as the ground truth for each observation.

3 Methodology

This work makes use of a standard machine learning pipeline as shown in Fig. 1. The following sub-sections discuss each individual component of the pipeline in further detail.

3.1 Data Pre-processing

Built on top the pre-processing done in [12], further processing was applied to the dataset for this work that included applying min–max normalisation of all continuous variables, scaling them to between 0 and 1 to remove any feature domination issues caused by discrepancies in ranges [22] and to bring them into a normal distribution, allowing them to be considered equally. Discrete data with a cardinality of 2 were scaled to 0 and 1 values for use as binary features, while those with high cardinality (>2) were one-hot encoded to separate the categories into independent features. The use of one-hot encoding removes ordinality implications between values that could affect results when considering distances.

3.2 Modelling Approach

The processed data was then split randomly into stratified k-folds ($k = 10$), maintaining the class ratios between each fold after which $k - 1$ folds were collated for use as training data for model development and the remaining fold kept separate for evaluating each trained instance. Given the imbalance between each of the data classes caused primarily by the limited number of AD samples, the training sets are modified to include generated observations created using the Synthetic Minority Oversampling TEchnique (SMOTE) to reduce bias towards the majority class [23]. The k-fold cross-validation process is repeated five times with randomised splits of the data upon each iteration to get a better estimate of the performance of the models across different combinations of data.

Feature Selection (FS) is applied as part of the pipeline to reduce the dimensionality of the data to sets that provide the most relevant information with respect to the classification task. Selection occurs for each k-th training set to reduce noise within the data [24], improving generalisation of model instances and computational efficiency [12]. For each set of derived features, the SOF model is developed using the oversampled training set and then evaluated against the left-out test fold. The results across all model evaluations are collated according to the metrics described in the evaluation section.

3.3 Feature Selection

While previous works considered univariate FS approaches, their use only considers how much a single variable can discriminate between prediction classes in isolation, leaving room for features to be removed that may provide greater value when used in combination with others. This work considers three FS options including no FS, as a base comparison; Binary Particle Swarm Optimisation (BPSO) [25, 26]; and Recursive Feature Elimination (RFE) [27], with the latter two being examples of multi-variate selection methods. BPSO and RFE are wrapper techniques that determine a final feature subset by building and evaluating models separately from the rest of the pipeline, returning the subset that produces the best evaluation measure across the multiple sets that are attempted. The use of wrapper methods is computationally more expensive than filter methods; however, they are generally known to produce better results through the use of actual model evaluation [26].

BPSO works primarily on the same basis as the popular Particle Swarm Optimisation, which aims to optimise a problem by iteratively trying to improve a candidate solution with regard to a given measure of quality. It works by having a population of candidate solutions (dubbed particles), and moving these particles around in the search-space with each particle's movement influenced by its local best-known position, but also guided towards the best-known positions in the search-space, which are updated as better positions are found by other particles. In our case, each particle enables a certain feature subset out of original 66 selected features. The binary aspect of BPSO comes in the position updates by converting the computed velocities into probabilities using, for example, a sigmoid function. By comparing the feature probabilities to random numbers that are uniformly distributed between zero and one, the particle position is updated to include features where the random number is below the probability, and exclude/remove features otherwise. Evaluations are made for the new positions and the process continues until a threshold number of iterations has passed at which point the global best position, or feature set, is returned.

Unlike BPSO, RFE is not stochastic-based, but systematic in that it recursively considers the relative importance of features at each iteration. The process starts by building and evaluating a model based on all available features which are then ranked according to a measure of importance such as, in the case of Support Vector Machines (SVM), coefficients learned by the model during training. From the ranked features, the least useful is removed from the set and a new model instance built and evaluated from the result. The process continues until a minimum threshold or a single feature remains. Based on the evaluation scores of the model for each feature subset, the set that produced the best values over all iterations is returned for use in the final model.

As the SOF classifier has no concept of coefficients or feature importance it is currently not possible to use RFE coupled with SOF directly for FS, as there is no metric to decide on what to remove. As such, SVM with a linear kernel was selected for the model as it has been successfully utilised in research [27]. The use of SVM was expanded for use in both RFE and BPSO to keep FS comparison fair between the two. With respect to FS model development, training of the SVM used stratified k-fold ($k = 5$) cross-validation (CV) on the training set with the evaluation of the model in each case being based upon the accuracy of the predictions made on the held-out fold from the selection CV, averaged over all iterations.

3.4 Model Development

Model development is achieved using the SOF offline training mode described in previous sections. After preprocessing the data and applying feature selection, the reduced dimensional data is input into the classifier. Upon starting training, the dataset is split into the three independent sets based on whether an observation is CU, MCI or AD according to the dependent variable. Each class is considered independently. For CU data, any duplicate observations (limited by the selected features) are removed leaving only the unique samples from the training set. For each unique CU observation, the multi-modal density is calculated and these are added to a list in order of the least distance from the tail of the list, starting with the highest density seen in the subset. The CU observations producing a peak when considering the densities within the list are separated out to produce a set of centroids, to which all other listed observations are assigned based on which centroid they are closest to. This forms one or more CU data clouds from which a new cloud centroid is produced and a density computed. Based upon granularity parameter and the CU data points, neighbourhoods are determined that group nearby centroids and then reduce them to a single point by using the one with the highest density. These CU centroids become the prototypes that are used to compare the distance to previously unseen data for classification describing the strength of the class. This process is repeated two more times: once for the MCI subset of data points, and once for AD resulting in three separate lists of prototype groups.

4 Experimentation

The methods discussed above were implemented in Python 3 using modules provided by Scikit Learn and PySwarms among others to provide functionality for preprocessing, oversampling, cross-validation, FS and evaluation. The MATLAB engine for Python was used to integrate the SOF classifier built-in MATLAB (developed by [21]) into the Python pipeline. The results of the experiments as described in the evaluation section above are shown in Table 4.

Multiple metrics are collated from model evaluations in order to determine the effectiveness of both the FS methods and the performance of the SOF classifier in the given domain. The metrics considered include the multi-class accuracy; balanced accuracy, which considers the imbalance of data classes by using the macro average of per-class recall; per-class precision and macro-precision; per-class recall and macro-recall; and per class F1 and macro-F1 (harmonic mean) scores. Macro in context meaning that the metrics are calculated per class and the mean taken. Each metric is averaged over all data folds with their respective standard deviations provided where appropriate. These are all common metrics used for the evaluation of classification tasks and were selected for that reason [28]. The use of average F1 rather than F1 of averages was determined by considering recommendations made by [29]. Area under the ROC curve was also considered as a potential metric; however, due to the nature of the SOF classifier there are no thresholds available to compute probabilistic predictions, making the use of the metric unhelpful when only a single point is available.

Each of the above measures was computed without FS, with BPSO and with RFE for comparison. The use of static random seeds in the implementation allowed for reproducible results and for identical data sets to be produced for each experiment particularly with respect to the data splits created for repeated k-fold and SMOTE samples. This allows for direct comparison of selection techniques and an overall impression of the SOF classifier's performance.

With respect to FS, RFE-SVM proved to produce feature subsets that allowed the SOF classifier to attain the best results compared to BPSO and no FS by over 10% in both multi-class accuracy and F1 scores, while also improving on balanced accuracy. The results of BPSO proved surprising with an average feature set size of 50.6, compared to the much smaller average of 11 features selected by RFE. Examination of the BPSO particle positions over time appeared to show that particles got stuck exploring narrowly around a local optimum feature subset, suggesting

Table 1 Top 10 occurrences of features selected by RFE

Feature	Modality	Occurances
LDELTOTAL	CM	50
FAQ	CM	50
MMSE	CM	50
ADAS13	CM	50
AGE	Patient	49
Hippocampus	MRI	45
Temporal_Left	PET	42
PTRACCAT_2	Patient	34
TAU_ABETA	CSF	34
LIMMTOTAL	CM	20

that convergence of the particles happened much too early and resulted in limited exploration. This appears to be a known issue with BPSO as discussed by [26]. Multiple variations of BPSO parameters were attempted including increases of both the population of particles and maximum attempted iterations in addition to testing different values for cognitive, social and inertial weights. Results varied with the maximum accuracy seen around 0.8 for a single run of tenfold CV, however, the computational time and power required for each iteration increased to the point where the process was no longer viable. The results shown in the table are from using the default 20 particles over 100 iterations with cognitive/social factors of 0.5 and an inertia weighting of 0.8 producing evaluations worse than using no FS at all.

Additionally, by using the feature sets derived by RFE, commonly seen features were tabulated (see Table 1). The table shows how the optimal features span across the data modalities described previously. In particular, multiple neurocognitive assessment scores appear in every set (LDELTOTAL, FAQ, MMSE, ADAS13) suggesting that these particular tests are the most useful for classification while others (LIMM-TOTAL, MoCA and RAVLT trials) are used in less than half showing that, perhaps, some aspects of brain function are unnecessary, or less helpful for AD classification. Age also appears often which, as discussed previously, is expected due to the higher chance of the condition later in life. Additionally, single features are picked from the MRI (Hippocampus), PET (Temporal_Left) and CSF (TAU_ABETA) modalities.

In terms of classification, when using the reduced features determined by RFE, the SOF classifier in offline training mode has been shown to produce good results, with an average accuracy and balanced accuracy of 0.81 and 0.82, respectively, with an F1-score of 0.81, all averaged over five-repeated tenfold CV. The accuracy values suggest that the use of SMOTE may have successfully reduced the effect of class imbalance. Specifically, these results were achieved using an SOF granularity of 1, the lowest available value, which results in a lower number of prototypes being generated for prediction purposes. Any value above 1 resulted in worse evaluation scores which, as discussed by [21] suggests that the lower values may allow the

Table 2 Summed RFE-SVM based confusion matrix over all iterations

		Predicted			
		CU	MCI	AD	Total
Actual	CU	816	148	6	970
	MCI	193	812	85	1090
	AD	1	42	337	380
	Total	1010	1002	428	2440

Table 3 Precision and recall results of experiments using FS variations and the SOF classifier

Feature selec-tion	Avg. features	Precision				Recall			
		NC	MCI	AD	Avg	NC	MCI	AD	Avg
None	*	0.72 ± 0.07	0.72 ± 0.11	0.68 ± 0.13	0.71 ± 0.07	0.79 ± 0.11	0.57 ± 0.09	0.85 ± 0.13	0.74 ± 0.07
RFE	11	0.81 ± 0.08	0.82 ± 0.09	0.81 ± 0.14	**0.82 ± 0.07**	0.84 ± 0.09	0.74 ± 0.13	0.89 ± 0.13	**0.82 ± 0.07**
BPSO	50.6	0.73 ± 0.08	0.71 ± 0.09	0.65 ± 0.12	0.70 ± 0.06	0.79 ± 0.10	0.56 ± 0.13	0.86 ± 0.10	0.73 ± 0.06

model to generalise better. By using more prototypes, it could be that overfitting was occurring due to the extra complexity caused by introducing more points (and radii) into the prototype space, effectively tightening the equivalent of the decision boundaries.

Table 2 shows the confusion matrix generated across all folds when using RFE. In general, the classifier did particularly well when separating CU and AD patients with only seven misclassifications made between the two, only one of which resulted in a prediction of being unimpaired when the patient was positive AD. On the other hand, most of the errors occurred when attempting to distinguish between MCI and AD/NC. This may be in part due to MCI being an intermediate stage with overlapping features that make definitive separations difficult between the two main diagnosis classes Tables 3 and 4.

5 Conclusion and Future Work

This report considers the use of fuzzy systems, specifically the recent SOF classifier for use in predicting whether baseline observations for patients seen in the ADNI data indicate that they suffer from AD, MCI or are CU while using FS methods. SOF is shown to be appropriate and useful in the domain of AD prediction, though the results did not reach as high in accuracy or F1 as other works including [12]. While the results

Table 4 Accuracy and F1 results of experiments using FS variations and the SOF classifier

Feature selection	Avg. features	Avg. accuracy/Bal. accuracy	F1			
			NC	MCI	AD	Avg
None	*	0.70 ± 0.06/0.74 ± 0.07	0.75 ± 0.07	0.63 ± 0.08	0.75 ± 0.10	0.71 ± 0.06
RFE	11	**0.81 ± 0.07/0.82± 0.07**	0.82 ± 0.06	0.77 ± 0.09	0.84 ± 0.10	**0.81 ± 0.07**
BPSO	50.6	0.69 ± 0.06/0.73 ± 0.06	0.75 ± 0.07	0.61 ± 0.10	0.73 ± 0.09	0.70 ± 0.06

fell short, the methods used to achieve them are much more robust and close enough to warrant further consideration. In particular, given the low inaccuracy between the main CU and AD classes, the experiment could potentially be used independently or, to reduce the likelihood of false positives of CU/false negatives for AD, as part of an ensemble AD screening tool using multiple ML techniques.

The limitations of this experiment revolve primarily around the data. First, the experiment was limited by the features and patients selected, better results may be possible by making fuller use of the ADNI data by considering more of the features and patients available. Second, the class imbalance within the sourced data, particularly between CU and AD, can cause issues with classifications even though attempts were made to reduce its impact by oversampling and considering balanced accuracy. As such considering the use of data imputation such as advanced interpolation techniques [30] in future work rather than removal of incomplete samples may produce better results by retaining a better class balance or keep enough data to consider undersampling. Third, the removal of the CDRSB variable is likely to have had a negative impact on the results as it is an important factor for diagnosis. Therefore, repeating the experiments with the CDRSB variable available as an independent feature may improve the presented results in future works.

The tabulated FS results show how useful features are split over multiple modalities. However, as shown by the number of observations remaining in the provided dataset, limited numbers of patients have data available for all considered modalities with, for example, more having undertaken neurocognitive testing and less for MRI/PET/CSF. Future work may consider an ensemble system where models are developed per modality with the final decision being made based on the results of each model for which data is available. The use of such an ensemble could fill a gap in screening that would allow an overall prediction per patient for all modality data available per patient with confidences determined by the support available for a particular class for stratification.

Acknowledgements Data collection and sharing for this project was funded by the Alzheimer's Disease Neuroimaging Initiative (ADNI) (National Institutes of Health Grant U01 AG024904) and DOD ADNI (Department of Defense award number W81XWH-12-2-0012). ADNI is funded by the National Institute on Aging, the National Institute of Biomedical Imaging and Bioengineering, and through generous contributions from the following: AbbVie, Alzheimer's Association; Alzheimer's Drug Discovery Foundation; Araclon Biotech; BioClinica, Inc.; Biogen; Bristol-Myers Squibb Company; CereSpir, Inc.; Cogstate; Eisai Inc.; Elan Pharmaceuticals, Inc.; Eli Lilly and Company; EuroImmun; F. Hoffmann-La Roche Ltd and its affiliated company Genentech, Inc.; Fujirebio; GE Healthcare; IXICO Ltd.; Janssen Alzheimer Immunotherapy Research & Development, LLC.; Johnson & Johnson Pharmaceutical Research & Development LLC.; Lumosity; Lundbeck; Merck & Co., Inc.; Meso Scale Diagnostics, LLC.; NeuroRx Research; Neurotrack Technologies; Novartis Pharmaceuticals Corporation; Pfizer Inc.; Piramal Imaging; Servier; Takeda Pharmaceutical Company; and Transition Therapeutics. The Canadian Institutes of Health Research is providing funds to support ADNI clinical sites in Canada. Private sector contributions are facilitated by the Foundation for the National Institutes of Health (www.fnih.org). The grantee organization is the Northern California Institute for Research and Education, and the study is coordinated by the Alzheimer's Therapeutic Research Institute at the University of Southern California. ADNI data are disseminated by the Laboratory for NeuroImaging at the University of Southern California.

References

1. Alzheimer's Research UK. (2018). *Global prevalence*. https://www.dementiastatistics.org/statistics/global-prevalence/.
2. World Health Organisation. (2019). *Dementia*. https://www.who.int/news-room/fact-sheets/detail/dementia.
3. NHS. (2019). *New type of dementia identified—NHS* [online]. https://www.nhs.uk/news/neurology/new-type-dementia-identified/.
4. NHS. (2018). *Alzheimer's disease—symptoms—NHS* [online]. https://www.nhs.uk/conditions/alzheimers-disease/symptoms/.
5. Alzheimer's Research UK. (2018). *Prevalence by age in the UK*. https://www.dementiastatistics.org/statistics/prevalence-by-age-in-the-uk/.
6. Alzheimer's Society. (2019). *Drug treatments for Alzheimer's disease | Alzheimer's Society* [online]. https://www.alzheimers.org.uk/about-dementia/treatments/drugs/drug-treatments-alzheimers-disease.
7. Kour, H., Manhas, J., & Sharma, V. (2019). Evaluation of adaptive neuro-fuzzy inference system with artificial neural network and fuzzy logic in diagnosis of Alzheimer disease. In *2019 6th International Conference on Computing for Sustainable Global Development (INDIACom)* (pp. 1041–1046). IEEE. https://ieeexplore.ieee.org/document/8991423.
8. Zhu, F., Li, X., McGonigle, D., Tang, H., He, Z., Zhang, C., Hung, G. U., Chiu, P. Y., & Zhou, W. (2020). Analyze informant-based questionnaire for the early diagnosis of Senile Dementia using deep learning. *IEEE Journal of Translational Engineering in Health and Medicine, 8*.
9. Cahill, S., Clark, M., O'Connell, H., Lawlor, B., Coen, R. F., & Walsh, C. (2008). The attitudes and practices of general practitioners regarding dementia diagnosis in Ireland. *International Journal of Geriatric Psychiatry, 23*(7), 663–669 [online]. http://doi.wiley.com/10.1002/gps.1956
10. Bradford, A., Kunik, M. E., Schulz, P., Williams, S. P., & Singh, H. (2009). Missed and delayed diagnosis of dementia in primary care: Prevalence and contributing factors. *Alzheimer Disease and Associated Disorders, 23*(4), 306.
11. NHS. (2019). *Guide to NHS waiting times in England—NHS* [Online]. https://www.nhs.uk/using-the-nhs/nhs-services/hospitals/guide-to-nhs-waiting-times-in-england/

12. Bucholc, M., Ding, X., Wang, H., Glass, D. H., Wang, H., Prasad, G., et al. (2019). A practical computerized decision support system for predicting the severity of Alzheimer's disease of an individual. *Expert Systems with Applications, 130*, 157–171.
13. Lee, G. G. C., Huang, P. W., Xie, Y. R., & Pai, M. C. (2019). Classification of Alzheimer's disease, mild cognitive impairment, and cognitively normal based on neuropsychological data via supervised learning. In *IEEE Region 10 Annual International Conference, Proceedings/TENCON* (Vol. 2019-October, pp. 1808–1812). Institute of Electrical and Electronics Engineers Inc.
14. Er, F., Iscen, P., Sahin, S., Çinar, N., Karsidag, S., & Goularas, D. (2017). Distinguishing age-related cognitive decline from dementias: A study based on machine learning algorithms. *Journal of Clinical Neuroscience.*
15. Chen, T., Antoniou, G., Adamou, M., Tachmazidis, I., & Su, P. (2019). Automatic diagnosis of attention deficit hyperactivity disorder using machine learning. *Applied Artificial Intelligence.*
16. Chen, T., Su, P., Shang, C., & Shen, Q. (2018). Weighted fuzzy rules optimised by particle swarm for network intrusion detection. In *2018 IEEE International Conference on Fuzzy Systems* (pp. 1–7). IEEE.
17. Su, P., Shang, C., Chen, T., & Shen, Q. (2017). Exploiting data reliability and fuzzy clustering for journal ranking. *IEEE Transactions on Fuzzy Systems, 25*(5), 1306–1319.
18. Chen, T., Shang, C., Su, P., Keravnou-Papailiou, E., Zhao, Y., Antoniou, G., et al. (2020). A decision tree-initialised neuro-fuzzy approach for clinical decision support. *Artificial Intelligence in Medicine.*
19. Su, P., Chen, T., Xie, J., Ma, B., Qi, H., Liu, J., & Zhao, Y. (2020). A density and reliability guided aggregation for the assessment of vessels and nerve fibres tortuosity. *IEEE Access.*
20. Chen, T., Su, P., Shang, C., Hill, R., Zhang, H., & Shen, Q. (2019). Sentiment classification of drug reviews using fuzzy-rough feature selection. In *2019 IEEE International Conference on Fuzzy Systems* (pp. 1–6). IEEE.
21. Gu, X., & Angelov, P. P. (2018). Self-organising fuzzy logic classifier. *Information Sciences, 447*, 36–51.
22. Han, J., Pei, J., & Kamber, M. (2011). *Data mining: Concepts and techniques.* Elsevier.
23. Blagus, R., & Lusa, L. (2013). SMOTE for high-dimensional class-imbalanced data. *BMC Bioinformatics, 14*(1), 1–16.
24. Carlos Molina, L., Belanche, L., & Nebot, À. (2002). Feature selection algorithms: A survey and experimental evaluation. In *Proceedings—IEEE International Conference on Data Mining, ICDM* (pp. 306–313).
25. Kennedy, J., & Eberhart, R. C. (1997). Discrete binary version of the particle swarm algorithm. In *Proceedings of the IEEE International Conference on Systems, Man and Cybernetics* (Vol. 5, pp. 4104–4108). IEEE.
26. Too, J., Abdullah, A. R., & Mohd Saad, N. (2019). A new co-evolution binary particle swarm optimization with multiple inertia weight strategy for feature selection. *Informatics, 6*(2), 21.
27. Maldonado, S., Weber, R., & Famili, F. (2014). Feature selection for high-dimensional class-imbalanced data sets using Support Vector Machines. *Information Sciences, 286*, 228–246.
28. Hossin, M., & Sulaiman, M. (2015). A review on evaluation metrics for data classification evaluations. *International Journal of Data Mining & Knowledge Management Process, 5*(2), 1.
29. Opitz, J., & Burst, S. (2019). *Macro F1 and Macro F1* [Online]. http://arxiv.org/abs/1911.03347
30. Chen, T., Shang, C., Yang, J., Li, F., & Shen, Q. (2019). A new approach for transformation-based fuzzy rule interpolation. In *IEEE Transactions on Fuzzy Systems* [Online]. https://doi.org/10.1109/TFUZZ.2019.2949767.

Autism Spectrum Disorder Classification Using a Self-organising Fuzzy Classifier

Jonathan Stirling, Tianhua Chen, and Marios Adamou

Abstract Autism Spectrum Disorder (ASD) is a neurodevelopmental disorder that covers a range of symptoms such as impaired social skills and repetitive behaviours. The diagnosis of ASD in clinics is typically lengthy and cost-ineffective. Recent advances in machine learning could facilitate more efficient and effective detection of ASD. However, fuzzy systems, as a significant soft computing technique, have been sporadically applied in the diagnosis of ASD. This chapter, therefore, examines the use of a recently proposed self-organising fuzzy classifier with application to the "autism screening adult" data retrieved from a mobile application.

Keywords Autism spectrum disorder · ASD · Clinical decision support · Fuzzy systems · Self-organising fuzzy classifier · SOF

1 Introduction

Autism Spectrum Disorder (ASD) is a neurodevelopmental disorder characterised by pervasive difficulties in reciprocal social interaction, alongside the presence of strict repetitive interests and behaviours [1]; studies estimate that approximately 1% of the population may have an ASD [2].

Whilst much research in ASD focuses on the developmental period, it is recognised that ASD is a lifelong condition [3], which is sometimes not detected clinically until later life. Compensation strategies may contribute to this delay, in that learnt behaviour or higher levels of intelligence, and level of severity can successfully mask autistic symptoms [4].

J. Stirling · T. Chen (✉)
Department of Computer Science, School of Computing and Engineering, University of Huddersfield, Huddersfield, UK
e-mail: T.Chen@hud.ac.uk

M. Adamou
South West Yorkshire Partnership NHS Foundation Trust, Wakefield, UK
e-mail: marios.adamou@swyt.nhs.uk

© Springer Nature Switzerland AG 2021
J. Carter et al. (eds.), *Fuzzy Logic*,
https://doi.org/10.1007/978-3-030-66474-9_6

83

Taking heed from recent NICE guidelines for ASD in adults [5] the diagnosis of ASD in adulthood is made based on consensus of expert opinion. This opinion is reached after analysing clinical information consisting of detailed psychiatric history, observations of a person's interaction during the assessment, scores of validated diagnostic tools and collected history provided by a person who knows the individual. This process may take place over several appointments and claim over 240 min of clinical time.

Adding to the difficulty in making a correct diagnosis, the clinical presentation of a person with Autism can greatly overlap with other disorders, specifically, negative symptoms of schizophrenia [6]. Together with possible co-morbidity [7], the picture can become complex [8] with a tangible risk of false positives from standardised diagnostic assessments.

A combination of factors has led to the creation of lists of patients waiting for years for their turn to receive an appointment for a diagnostic assessment by the UK National Health Service. Providing early, fast and reliable methods of diagnosis is beneficial both economically and in long-term outcomes for patients. More generally, diagnosis can make a difference to quality of life with access to the relevant healthcare and community support, financial help and in some cases peace of mind when a diagnosis has been assumed then finally confirmed.

The popularity of Machine Learning (ML) and its successes in problem domains including those of health care is evident in the volume of papers being released year by year [9, 10]. This is in part by the nature of ML that it can make use of the vast quantity of data maintained by healthcare providers to provide insight and clinical support for a range of problems. This extends to the use of ML in ASD research which considers various different approaches, commonly in regard to binary diagnosis by considering various data sources and, commonly, supervised ML techniques [11].

As to be reviewed in the next section, SVM and tree-based ML techniques have been the most popular options for ASD prediction to date, with a number of neural network and other techniques being used occasionally according to a survey by [11]. An area that appears sporadically is the use of fuzzy systems such as works using neuro-fuzzy, merging fuzzy systems and NN [12].

As a significant soft computing technique, fuzzy rule-based systems [13] are constructed on the basis of fuzzy sets and fuzzy logic, which allows the representation and manipulation of information that are vague and lack certainty. This has led to fuzzy systems applied in various medical applications to deal with uncertainty and imprecision that may result from linguistic descriptions while enquiring medical symptoms or noise that may result from inaccurate testing results [14–17].

However, recent studies in this area appear limited in their scope and depth. Additionally, works detailed above most often concentrate on child diagnosis rather than those of adults where childhood diagnosis has been missed or not tested for. As such, in working towards providing assistance for clinicians to conduct effective ASD diagnosis, this work considers the use of the non-parametric Self-Organising Fuzzy (SOF) classifier developed by [18] to determine the applicability of the technique in the prediction of ASD in adults.

The remainder of this chapter is organised as follows. Section 2 reviews recent advances in ASD diagnosis. Section 3.1 through 3.6 describes the proposed pipeline. Section 3.7 presents and discusses the experimental outcomes. Section 4 concludes the chapter and outlines ideas for further development.

2   Literature Review

Current clinical practice makes use of a variety of ASD diagnostic instruments [19] including the Autism Diagnostic Observation Schedule Generic (ADOS), made up of four modules, each for use under specific conditions; the Autism Diagnostic Interview-Revised (ADI-R), made up of 93 items for individuals mentally over 18 months old; and Autism Spectrum Quotient (AQ) tests, used for adult diagnoses and made up of 50 items though variations were created to cover younger subjects [20]. The results of these instruments, or most often combinations of them, are coupled with deliberations by a team of clinicians to agree on a Best Estimate Clinical diagnosis of whether a patient has ASD [21].

Clinical instruments used for diagnosis cover a range of symptoms that can appear in various combinations and levels where such manifestations of behaviours can inform a diagnosis [22]. Being able to recognise these aspects via technology has allowed for classifications to be attempted independently of existing instruments by extracting specific behavioural measures from, e.g. sensors/video data and then comparing between ASD and Typical Development (TD) controls. Of these behaviours, recent studies single out variations in speech, eye-movement and motor skills for consideration as well-established symptoms of ASD.

Nakai et al. [23] and Lee et al. [24] consider using features extracted from single word audio recordings as a predictor for ASD. Nakai et al. [23] extracted 24 features from the recordings, and along with clinical diagnoses made by paediatric neurologists built SVM prediction models that gave positive results (73% accuracy; 0.76 f-score). Similarly, INTERSPEECH 2013 Computational Paralinguistics Challenge participants were asked to classify child observations into binary and multi-class labels using provided utterances. For this challenge Lee et al. [24] expands beyond singular techniques demonstrating an ensemble ML system using SVM, deep neural networks and a weighted discrete version of K-Nearest Neighbour (KNN) models along with an "acoustic segment model" that classifies temporal features extracted from audio. The weighted sum of posterior probabilities across the four classifiers determined the final result, and using unweighted average recall as a performance metric, achieved 92.2% and 64.8% for binary and multi-class predictions respectively. In addition to classifier performance measures, Nakai et al. [23] also compared their results with those from trained speech therapists and show a definite improvement in using these techniques over human hearing specifically in the instance of single utterance classification.

Eye-movement was considered by [25] and [26] by using tools more often seen in Natural Language Processing, such as N-gram and Bag of Words to represent participants' gaze and eye-movement vector data as they are presented with a sequence of faces. The resulting features were then used to train an SVM using a radial-basis function kernel as a binary classifier for gaze and motion features independently, then combined, to classify participants' per-face inputs as ASD or not. Threshold tests were then applied on the mean of classification values over all displayed faces to determine participant-level diagnosis. Liu et al. [25] considered both adult and child independently with results of 0.7277/0.8033 (AUC/Accuracy) in adults and 0.9208/0.8689 (AUC/Accuracy) in children. Liu et al. [26] expands upon the child classifiers with more specific datasets confirming previous results but also finding that the use of faces of people from races other than the participant's resulted in better performance, which may be useful in further studies.

Research also considers movement, or kinematics, to measure motor features of participants [11]. Two recent variations on this area consist of Li et al. [22] asking participants to imitate the motions shown to them in videos using motion sensors to track 20 kinematic features, and Crippa et al. [27] who use to demonstrate two-movement reach-grasp-position-drop tasks to extract 17 features from sensors. With respect to techniques, Crippa et al. [27] use SVM with a linear kernel while Li et al. [22] considered both linear and radial-basis function kernels as well as Naïve Bayes for the classification task. Each work uses feature selection to reduce the number of dimensions to the most useful. In terms of selection Crippa et al. [27] used the Fisher discriminant ratio resulting in a combination of 7 discriminating features from the original 17 that allowed the SVM model to achieve a best accuracy of 96.7%. Li et al. [22] instead attempted to use multiple selectors including the SVM weights, PCA weights and Leave-one-parameter-out methods, determining the best classifier and selector combination for their data to be SVM (with linear kernel), with feature selection determined by a combination of the three mentioned options giving a best accuracy of 86.7%.

ASD is a neurodevelopmental disorder, thus the consideration of directly extracting metrics of brain function which translate into specific behaviour is of interest both in making diagnoses and in attempting to provide greater understanding of the physical effects of ASD on the brain. Bosl et al. [28] use recordings from an EEG for children aged 6–24 months, stipulating that the recorded signals can be used as biomarkers for distinguishing normal brain development and irregular development using MultiScale Entropy (MSE), while Zhang et al. [29] have considered how white matter impairments tracked across the whole brain may be used as potential biomarkers for ASD prediction through diffusion MRI (dMRI) scans from 6 to 18 year-olds. Zhang et al. [29] used feature extraction of valid fibre clusters followed by the selection of those clusters that provide the most information in diagnosing ASD using a signal-to-noise ratio coefficient before training an SVM model. Similarly, Bosl et al. [28] consider SVM but add KNN and Naïve Bayes techniques to find the best for classification using the means of MSE (mMSE) as feature vectors. With respect to results, the use of dMRI and mMSE showed promise, particularly for males with near 100% accuracy at 9 months of age and girls with 80% accuracy at 6 months when using SVM, compared to Zhang et al. [29] who realised a 78.33% best accuracy when considering 4697 fibre clusters. While the use of dMRI looks particularly promising, especially when considering early diagnosis, it has the same problems as current methods in terms of cost, professional staff requirements and waiting times for scans perhaps limiting its usefulness.

3 Materials and Methods

3.1 Participants

This work considers the use of the "Autism Screening Adult" data retrieved from a mobile application developed by [20] and available in UCI Machine Learning Repository [30] which, for adult patients, asks ten questions from the AQ series (AQ-10) that are used for screening purposes. The questions can be answered by anybody including the patient, or somebody else on their behalf, where the person using the app is considered the "user". The full set consists of 704 observations made up of 20 independent variables and one categorical dependent variable (*Class/ASD*). Of the independent variables ten are binary, representing positive or negative responses to each question where a value of one indicates an expected increase in the likelihood of ASD (*A{1-10}_Score*). Four variables are categorical, covering a variety of demographic data including the patient's ethnicity (*ethnicity*); country of residence (*country_of_res*); and age group (*age_desc*). A further four categorical variables cover whether the patient was born with jaundice (*jaundice*); a member of their family has a pervasive developmental disorder (*autism*); the user has previously used the mobile app (*used_app_before*); and who the user is in relation to the patient (*relation*), for example, "Self" or "Parent". The remaining two variables are continuous, describing the age (*age*; 29.7 ± 16.51) and summed total of question results (*result*; 4.88 ± 2.5). The dependant variable provided contains only "YES" and "NO" values for whether the observed person is predicted to suffer from ASD, making this a binary classification problem. Additionally, the observations are unbalanced over the positive and negative classes with a ratio of 189 positive and 515 negative instances.

3.2 Data Pre-processing

Before applying feature selection to the dataset, pre-processing is a necessary step to clean up the data to be usable by ML techniques and in a format that allows for their best performance. With respect to cleaning, the dataset contains 192 missing values spread across 95 observations and over three variables (*age, ethnicity* and *relation*) which for simplicity were removed from the set rather than attempting to use advanced imputation techniques such as [31]. The *age_desc* variable is removed from the dataset as all values within the variable are the same across all observations and similarly the *used_app_before* variable is also removed as this is irrelevant to the classification and not guaranteed to relate to the patient. The continuous variables (*age*, *result*) were min-max scaled to the range of zero to one with a standard deviation of one, while the nominal variables were label encoded to integer values, then one-hot encoded for those variables with a cardinality greater than two. This encoding removes issues with ordering/hierarchical inferences where no such relationships exist. For example, considering country of residence values of "United States", "New Zealand" and "Italy", there is no ordinal relationship between the values, they are equally as different from each other; however, using integer encoding alone would consider "United States" and "Italy" as being incorrectly further apart in context than "United States" and "New Zealand". Finally, the original feature and the first created feature of each one-hot encoding, respectively, are removed as the information has been encoded fully in the remaining features.

3.3   Feature Selection and Modelling Approach

A standard ML pipeline is implemented in this work following the process shown in Fig. 1. First, pre-processing of the data occurs to prepare the dataset for use as

Fig. 1 ML pipeline

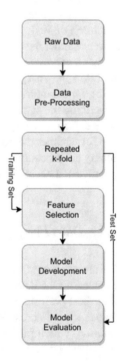

described above. Once complete, the result is split into stratified k-fold (k = 10) where the k-th fold is kept separate for use as a testing set while the remaining $k - 1$ folds are collated to form the training set. The training set is then oversampled using the Synthetic Minority Oversampling Technique to address the previously mentioned class imbalance for model development by generating new samples based upon the values from the originals of the minority class. Using k-fold cross validation guarantees that all observations in the dataset appear once in the testing set, and by averaging the results across all ten splits of data a more accurate measure of the model performance is obtained such that any bias between the splits is reduced. The k-fold process is repeated five times to further reduce any bias through repeated data shuffling between each run of ten folds. For every fold, feature selection is used on the training set to reduce the dimensionality of the data used to build the model, providing only the most useful items pertaining to the classification task. Once a feature set has been generated, the oversampled training set is used to develop the classification model, restricting the inputs to only the features that were selected. After training, the model is then evaluated using the testing set that was held out which, since the model has never seen the data previously, gives an idea of how well it can generalise.

3.4 Feature Selection

Two feature selection methods were implemented independently, consisting of a univariate filter method and a wrapper method. Such selection techniques reduce the total number of features used as inputs for classification models where noisy or irrelevant data can cause overfitting and/or negatively impact performance. Wrapper methods make use of ML techniques to develop models that make predictions based on subsets of features where the best evaluation found across all tested feature sets is used to develop the final classifier [32]. Filter methods by contrast do not rely on additional classifiers or techniques, but rather use statistical measures of the data to determine how important features may be for prediction [32].

The Recursive Feature Elimination (RFE) technique was selected as the wrapper method for its simplicity and ease of implementation. SVM with a linear kernel was used as the internal classifier to determine the best feature set using the average accuracy from models developed with stratified tenfold cross validation over the training set. The coefficients from the training model are considered as a ranking of the importance of each of the features in the tested set and used to remove the least important feature(s) recursively until only one remains. SOF was not used as the classifier for RFE as it does not provide a metric that can be used to rank features, which is required for this method. The set that produces the best evaluation or fitness, which in this case considers the accuracy of the model, is then selected for development.

For filter selection, an analysis of the information gain/mutual information of the features against the dependent variable is made where information gain effectively describes how well any one feature can be used to split the data samples between each of the data classes. The analysis suggests possible important features with respect to classification; however, given it only considers individual features, optimal sets are undetermined. For this work, the top 5% of features ordered by information gain are selected for further development.

3.5   Modelling

With respect to the model an SOF classifier [18] is introduced to confirm its viability for use in binary classification tasks. The classifier is a prototype-based system that creates single compressed AnYa type rule separately for every class in the data, describing the strength of a particular observation to each rule's class. The rules are initially generated through offline training, with online training also available to further train the model on new data points, however, only offline training is considered in this instance to provide a baseline for the model's performance and to not further reduce the dataset.

Training the model involves the creation of prototypes which are calculated points within the data space that are representative of areas of relevance per class. To generate these prototypes the multi-modal densities of all unique points for a single class are calculated and used to generate a ranking starting with the highest density point, then recursively adding the point closest in distance to the previously added item on the list. The points in the list where a density peak occurs (highest density between the adjacent points in the list) become pseudo-prototypes. These are used to form data clouds by assigning every other point to the cloud of its closest prototype. The centres of each cloud are calculated along with their multi-modal density, after which close neighbours of each cloud are identified. Finally, the cloud centres with the highest density of those in each neighbourhood are selected as the true prototypes and used to create the fuzzy rule. This process occurs for all classes with no interaction or interference between each instance. To make predictions, the distance to each prototype from a new data point is calculated, and the minimum distance found for each class is used as the strength of the class (where the lower the distance, the stronger the class). The minimum distance between all classes is then used as the final prediction. For this work the distances used are computed in euclidean space.

3.6   Evaluation

Evaluation of the developed model will consider averaged metrics calculated across each repeated k-fold iteration to reduce any bias that may be introduced by the use of specific data inputs, while providing an overall performance assessment of the model. Evaluation will be applied for both feature selection techniques separately for comparison purposes.

The metrics selected are those that are seen most commonly used in ML classification reporting including the overall accuracy of the model, balanced accuracy (added for comparison to account for unbalanced classes), precision and recall. The precision and recall values are further used to calculate the harmonic mean of the two (F1-score).

3.7 Results

The results of the experiment are summarised in Table 1. These results raise a number of concerns between the use of the dataset provided and the classifier under test. First, prior statistical analysis showed that the dependent variable was calculated automatically using a threshold of a single variable that is included within the dataset as an independent feature. Specifically, the *result* feature is the sum of the binary scores from the ten questions that make up AQ-10 where a value of one is applied if the selected answer is indicative of possible autism. Specifically, all values of *result* greater than six appear as "YES" for ASD, otherwise "NO". While this would imply that the use of such a dataset in a ML pipeline is moot, the findings did raise some important points.

First, filter feature selection methods are based upon statistical measures of individual features not on true model performance, and nor do they consider importance of combinations of features. As such, their use in an automated pipeline requires the use of thresholds that attempt to select a helpful set. As shown by the results using an automated filter method selecting 5% of the ranked features, the model only achieved 93% accuracy with precision being particularly affected, while standard deviations show a relatively large spread of individual results even while the *result* variable is included. While an important option in terms of selection techniques, without further evaluation automatic filter methods may not produce optimal results, especially in cases where some features within the selection threshold are actually not important for classification. Additionally, this shows how noisy data can lead to bad performance.

Table 1 Results of SOF model on testing set averaged over 5 repeated tenfold cross validation

Feature selection	Accuracy	Balanced accuracy	Precision	Recall	F1-score
Filter univariate	0.93 ± 0.04	0.93 ± 0.04	0.85 ± 0.09	0.93 ± 0.07	0.89 ± 0.07
RFE-SVM (Original)	0.90 ± 0.30	0.90 ± 0.30	0.90 ± 0.30	0.90 ± 0.30	0.90 ± 0.30
RFE-SVM (Adjusted)	1	1	1	1	1

The wrapper method on the other hand consistently selected the single variable (*result*) feature set by using an SVM classifier which used classification accuracy as the guide for selection. However, use of this method was far more computationally expensive by the nature of having to train and evaluate models multiple times. With the use of the single feature, problems started occurring with the classifier algorithm when training against a small number of attempted training sets. Specifically, when creating a ranking of data points according to distance to the latest point in the ranking, in particular when there are limited unique values available across features, no items in the list created a peak in multi-modal density causing the algorithm to only select the maximum density for consideration. By only having a single point to consider, no further distances were computed as there was nothing to compare to, resulting in no prototypes being generated for, in this case, the positive class. During testing, the SOF prediction process, therefore, failed due to missing any prototypes for the class. The "original" row in the results table shows the average metrics over all folds by assuming that each failed iteration made zero correct predictions. By adjusting the algorithm to force the single point to be used as a pseudo-prototype when no others are available the classifier worked as expected, returning the values shown in the "adjusted" row of Table 1.

4   Conclusion

This chapter considers the use of the "Autism Screening Adult" dataset from UCI and an SOF classifier for making predictions of whether an individual should be considered as more likely to have ASD, and therefore a higher priority for further testing and diagnosis. It was determined that the dataset has limited value as a classification task for ML techniques when all that is required is a threshold test based on self-reported answers to questions; however, it is useful for displaying how differences in methods can affect the performance of classification models. Additionally, a problem with the SOF algorithm was found that caused errors but was resolved with a small change that resulted in the expected perfect accuracy shown in the results. Also, while the scope of the work was limited, the classifier did prove to give good results with the inclusion of noisy data and proved that a threshold function can be modelled with expected results.

Future works for assisting ASD diagnoses should potentially not use this dataset for anything beyond ML demonstrations unless a clinical diagnosis can be added to its observations. The additional data would allow for confirmation of whether AQ-10, with the provided demographic features, can give sufficient information for accurately predicting ASD through ML techniques. Otherwise, other datasets should be considered for use. Additionally, while the chapter shows that the SOF classifier can be used for prediction in this instance, further work is required to confirm the viability of it in more complex cases, as well as validation of the algorithm to remove further edge case issues that may be present.

References

1. Lord, C., Risi, S., Lambrecht, L., Cook, E. H., Leventhal, B. L., Dilavore, P. C., et al. (2000). The autism diagnostic observation schedule-generic: A standard measure of social and communication deficits associated with the spectrum of autism. *Journal of Autism and Developmental Disorders, 30*(3), 205–223.
2. Baron-Cohen, S., Scott, F. J., Allison, C., Williams, J., Bolton, P., Matthews, F. E., et al. (2009). Prevalence of autism-spectrum conditions: UK school-based population study. *The British Journal of Psychiatry, 194*(6), 500–509.
3. Murphy, C. M., Wilson, C. E., Robertson, D. M., Ecker, C., Daly, E. M., Hammond, N., et al. (2016). Autism spectrum disorder in adults: Diagnosis, management, and health services development. *Neuropsychiatric Disease and Treatment.*
4. Bastiaansen, J. A., Meffert, H., Hein, S., Huizinga, P., Ketelaars, C., Pijnenborg, M., et al. (2011). Diagnosing autism spectrum disorders in adults: The use of autism diagnostic observation schedule (ADOS) module 4. *Journal of Autism and Developmental Disorders, 41*(9), 1256–1266.
5. National Institute for Health and Care Excellence. (2016). Autism spectrum disorder in adults: Diagnosis and management (guideline CG142).
6. Barlati, S., Deste, G., Gregorelli, M., & Vita, A. (2019). Autistic traits in a sample of adult patients with schizophrenia: Prevalence and correlates. *Psychological Medicine, 49*(1), 140–148.
7. Leyfer, O. T., Folstein, S. E., Bacalman, S., Davis, N. O., Dinh, E., Morgan, J., et al. (2006). Comorbid psychiatric disorders in children with autism: Interview development and rates of disorders. *Journal of Autism and Developmental Disorders, 36*(7), 849–861.
8. Molloy, C. A., Murray, D. S., Akers, R., Mitchell, T., & Manning-Courtney, P. (2011). Use of the autism diagnostic observation schedule (ADOS) in a clinical setting. *Autism, 15*(2), 143–162.
9. Chen, T., Antoniou, G., Adamou, M., Tachmazidis, I., & Su, P. (2019). Automatic diagnosis of attention deficit hyperactivity disorder using machine learning. *Applied Artificial Intelligence.*
10. Chen, T., Su, P., Shang, C., & Shen, Q. (2018). Weighted fuzzy rules optimised by particle swarm for network intrusion detection. In *2018 IEEE International Conference on Fuzzy Systems* (pp. 1–7). IEEE.
11. Hyde, K. K., Novack, M. N., LaHaye, N., Parlett-Pelleriti, C., Anden, R., Dixon, D. R., & Linstead, E. (2019). Applications of supervised machine learning in autism spectrum disorder research: A review. *Review Journal of Autism and Developmental Disorders, 6*(2), 128–146.
12. Arthi, K., & Tamilarasi, A. (2008). Prediction of autistic disorder using neuro fuzzy system by applying ANN technique. *International Journal of Developmental Neuroscience, 26*(7), 699–704.
13. Chen, T., Shang, C., Su, P., & Shen, Q. (2018). Induction of accurate and interpretable fuzzy rules from preliminary crisp representation. *Knowledge-Based Systems, 146*, 152–166.
14. Su, P., Chen, T., Xie, J., Zheng, Y., Qi, H., Borroni, D., et al. (2020). Corneal nerve tortuosity grading via ordered weighted averaging-based feature extraction. *Medical Physics.*
15. Chen, T., Su, P., Shang, C., Hill, R., Zhang, H., & Shen, Q. (2019). Sentiment classification of drug reviews using fuzzy-rough feature selection. In *2019 IEEE International Conference on Fuzzy Systems* (pp. 1–6). IEEE.
16. Su, P., Chen, T., Xie, J., Ma, B., Qi, H., Liu, J., & Zhao, Y. (2020). A density and reliability guided aggregation for the assessment of vessels and nerve fibres tortuosity. *IEEE Access.*
17. Chen, T., Shang, C., Su, P., Keravnou-Papailiou, E., Zhao, Y., Antoniou, G., et al. (2020). A decision tree-initialised neuro-fuzzy approach for clinical decision support. *Artificial Intelligence in Medicine.*
18. Gu, X., & Angelov, P. P. (2018). Self-organising fuzzy logic classifier. *Information Sciences, 447*, 36–51.
19. Lord, C., Corsello, C., & Grzadzinski, R. (2014). Diagnostic instruments in autistic spectrum disorders. *Handbook of autism and pervasive developmental disorders* (4th ed.).

20. Thabtah, F. (2019). An accessible and efficient autism screening method for behavioural data and predictive analyses. *Health Informatics Journal, 25*(4), 1739–1755.
21. Kassraian-Fard, P., Matthis, C., Balsters, J. H., Maathuis, M. H., & Wenderoth, N. (2016). Promises, pitfalls, and basic guidelines for applying machine learning classifiers to psychiatric imaging data, with autism as an example. *Frontiers in Psychiatry, 7*(DEC), 177.
22. Li, B., Sharma, A., Meng, J., Purushwalkam, S., & Gowen, E. (2017). Applying machine learning to identify autistic adults using imitation: An exploratory study. *PLoS ONE, 12*(8).
23. Nakai, Y., Takiguchi, T., Matsui, G., Yamaoka, N., & Takada, S. (2017). Detecting abnormal voice prosody through single-word utterances in children with autism spectrum disorders. *Perceptual and Motor Skills, 124*(5), 31512517716855.
24. Lee, H.-y., Hu, T.-y., Jing, H., Chang, Y.-F., Tsao, Y., Kao, Y.-C., & Pao, T.-L. (2013). Ensemble of machine learning and acoustic segment model techniques for speech emotion and autism spectrum disorders recognition. In *INTERSPEECH* [Online]. https://www.isca-speech.org/archive/archive_papers/interspeech_2013/i13_0215.pdf.
25. Liu, W., Yi, L., Yu, Z., Zou, X., Raj, B., & Li, M. (2015). Efficient autism spectrum disorder prediction with eye movement: A machine learning framework. In *2015 International Conference on Affective Computing and Intelligent Interaction, ACII 2015* (pp. 649–655). Institute of Electrical and Electronics Engineers Inc.
26. Liu, W., Li, M., & Yi, L. (2016). Identifying children with autism spectrum disorder based on their face processing abnormality: A machine learning framework. *Autism Research, 9*(8), 888–898 [Online]. https://doi.org/10.1002/aur.1615.
27. Crippa, A., Salvatore, C., Perego, P., Forti, S., Nobile, M., Molteni, M., et al. (2015). Use of machine learning to identify children with autism and their motor abnormalities. *Journal of Autism and Developmental Disorders, 45*(7), 2146–2156.
28. Bosl, W., Tierney, A., Tager-Flusberg, H., & Nelson, C. (2011). EEG complexity as a biomarker for autism spectrum disorder risk. *BMC Medicine, 9*(1), 18.
29. Zhang, F., Savadjiev, P., Cai, W., Song, Y., Rathi, Y., Tunç, B., et al. (2018). Whole brain white matter connectivity analysis using machine learning: An application to autism. *NeuroImage, 172*, 826–837.
30. UCI. (2017). UCI machine learning repository: Autism screening adult data set [Online]. https://archive.ics.uci.edu/ml/datasets/Autism+Screening+Adult.
31. Chen, T., Shang, C., Yang, J., Li, F., Shen, Q. (2019). A new approach for transformation-based fuzzy rule interpolation. *IEEE Transactions on Fuzzy Systems* [Online]. https://doi.org/10.1109/TFUZZ.2019.2949767.
32. Hall, M. A., & Smith, L. A. (1999) Feature selection for machine learning: Comparing a correlation-based filter approach to the wrapper. In *FLAIRS Conference* (Vol. 1999, pp. 235–239) [Online]. https://www.aaai.org/Library/FLAIRS/1999/flairs99-042.php.

An Outlier Detection Informed Aggregation Approach for Group Decision-Making

Chunru Chen, Tianhua Chen, Zhongmin Wang, Yanping Chen, and Hengshan Zhang

Abstract In group decision-making, owing to differences that may result from per-spectives such as experience and knowledge, the evaluations about the same decision problem provided by different crowd participants may have great differences. Those with huge differences in evaluations from most participants are termed outliers in this chapter. Reaching a decision consensus that satisfies most people is very difficult. In order to solve this problem, many researchers have conducted consensus research. To avoid this problem, this chapter proposes an expert opinions aggregation method based on outlier detection. First, the decision-maker evaluates the decision prob-lem based on the Pythagorean Fuzzy Sets (PFSs) from the positive and the negative views. Second, the outliers of expert opinions are detected and then aggregated to obtain the overall decision result. The effectiveness of the proposed method is finally demonstrated using a case study.

The original version of this chapter was revised: The name of the co-author has been corrected from "Tianghua Chen" to "Tianhua Chen". The correction to this chapter is available at https://doi.org/10.1007/978-3-030-66474-9_16

C. Chen · Z. Wang · Y. Chen · H. Zhang (✉)
School of Computer Science, Xian University of Post and Telecommunication,
Xi'an, China
e-mail: zhanghs@xupt.edu.cn

Z. Wang
e-mail: zmwang@xupt.edu.cn

Y. Chen
e-mail: chenyp@xupt.edu.cn

T. Chen
Department of Computer Science, School of Computing and Engineering,
University of Huddersfield, Huddersfield, UK
e-mail: T.Chen@hud.ac.uk

C. Chen · Z. Wang · Y. Chen · H. Zhang
Shanxi Key Laboratory of Network Data Analysis and Intelligent processing, Xi'an, China

Keywords Group decision-making · Consensus measure · Outliers · Pythagoras fuzzy set

1 Introduction

Decision-making is an indispensable activity in human activities, and it has an important role in solving practical problems in the political, military, and economic fields. In many practical applications, the decision-making problem may be very complicated, and it needs to rely on the opinions of a panel of experts in the field. By aggregating the opinions of involved experts, more accurate and reasonable decision-making results may be obtained. For example, in order to improve the operational effectiveness, mobile phone manufacturers must identify the needs of young users from various aspects of mobile phone performance. By aggregating the evaluations of users, the order or weight of mobile phone quality indexes can be made more reasonable.

Due to the complexity of decision-making problems, participants unfamiliar with decision-making problems, or different expressions of the problems, will lead to the opinions expressed by decision-makers being inaccurate or unavailable for calculation. Generally speaking, there are four ways in which the experts can express their opinions: preference orderings, utility values, fuzzy preference relations, and multiplicative preference relations [1]. Fuzzy sets, which generalize classical binary sets, have been used in a wide range of domains where information is incomplete or imprecise [2–4]. As experts understand the actual meanings of alternatives, which is bound to be uncertain, fuzzy sets can well express the imprecise and subjective opinions in pairwise comparison [5]. Pythagorean Fuzzy Sets (PFSs) are a generalization of intuitionistic fuzzy sets, in which decision-makers use linguistic scale close to natural language to express the decision results of the decision-makers. According to the actual meaning of alternatives, decision-makers can express their opinions on the importance of comparison between alternatives from the positive and negative views, and reflect the uncertainty of comparison in certain alternatives.

In group decision-making (GDM) with PFSs, there are two measures that have been considered before obtaining a final solution [6–8]: individual consistency and consensus. The individual consistency is applied to ensure that decision-maker is being neither random nor illogical in his/her pairwise comparisons. And the consensus ensures that the group of decision-makers agree on the final result to some extend [9–11]. And the individual consistency improving process is applied before the consensus reaching process. Preference relationships help to improve the individual consistency and avoid illogical and conflicting evaluations. In order to improve the individual consistency and consensus, many scholars have conducted substantial research and achieved fruitful results [9, 12–16]. In [9], the algorithm is also used to improve individual consistency and consensus simultaneously. Zhang et al. proposed an optimization model [12] to improve the individual consistency and consensus. But these methods generally require modifying the original opinions of decision-makers and can be very time-consuming in the solution process. In case there exists a large number of alternatives, it is very difficult to solve the optimal model. In [13], Zhang et al. proposed the method to improve the linguistic pairwise comparison consistency via linguistic discrete regions without modifying the original elements. And the method of crowd intelligence is proposed in [14]. When the decision-maker's opinion meets the threshold of consistency and consensus, the expert opinion is aggregated to obtain the weight or order of expert opinions[16].

Based on the above analysis, this chapter proposes an outlier detection informed method to aggregate decision-makers' evaluations. Decision-makers use the PFSs to give the pairwise comparison evaluations of alternatives from the positive and negative point of views, avoiding the problem of individual consistency. Then, for each alternative, with the detection of outliers, the consensus is aggregated following a Mixed Aggregation (MA) strategy. It is worth noting that directly detecting and improving consistency and consensus becomes practically difficult due to the recent advancement of large-scale group decision-making. The proposed method in this chapter, however, enables to avoid the consistency and consensus issue without needing to modify or disregard original opinions from decision-makers.

2 Preliminaries

GDM usually includes the following steps: (1) The panel's experts express the evaluations of alternative; (2) Detect and improve the individual consistency; (3) Detect and improve the consensus; (4) Aggregate the opinions of decision-makers and obtain the weight or order of the alternatives. According to [17], the decision-maker compares the alternatives from the positive and the negative views based on the PFSs, thereby avoiding the individual consistency. According to the evaluations provided by the decision-makers, the outliers of the evaluations significantly different from alternatives are detected, and the corresponding aggregation method is proposed to aggregate the opinions, which reduces the influence of the outliers and thus improves the consensus. Therefore, the method in this chapter does not need to detect and improve the individual consistency and consensus, and thus there is no need to modify the opinions of decision-makers.

2.1 Pairwise Comparisons of Alternatives Based on PFSs from the Positive and Negative Views

A GDM problem can be described as a group of decision-makers $E = \{e_1, e_2, \ldots e_n\}$ to make an order for a set of alternatives $A = \{A_1, A_2, \ldots A_n\}$. Under the scenario of evaluating mobile phone quality [18], let $E = \{e_1, e_2, \ldots e_n\}$ denote the decision-makers, $A = \{A_1, A_2, A_3\}$ be the indexes, which correspond to "Memory size of mobile phone", "Mobile screen resolution", "Cell camera pixels", respectively. And $S = \{s_1, s_2, \ldots, s_g\}$ be the linguistic term set. In the following, the utilized linguistic term set is $\{s_0 =$ extremely unimportant, $s_1 =$ very unimportant, $s_2 =$ less unimportant, $s_3 =$ unimportant, $s_4 =$ equal important, $s_5 =$ important, $s_6 =$ less important, $s_7 =$ very important, $s_8 =$ extremely important$\}$. The evaluations of decision-makers are presented by $M = \left(m_{ij} \right)_{n \times n}$, where $m_{ij} = \left(s_{ij}^u, s_{ij}^v \right)$, and the s_{ij}^u is the comparing result of decision-maker on A_i over A_j from the positive view which is called member degree, and s_{ij}^v is the one from the negative view which is called non-member degree. For instance, a decision-maker can express the evaluations using the following matrix:

$$D_1 = \begin{bmatrix} (s_4, s_4) & (s_8, s_1) & (s_7, s_2) \\ (s_0, s_7) & (s_4, s_4) & (s_3, s_6) \\ (s_1, s_6) & (s_5, s_2) & (s_4, s_4) \end{bmatrix}$$

After the decision-makers give their opinions, these opinions are translated into the Pythagoras Fuzzy Numbers (PFNs). First, translate the decision-maker's view into a numerical matrix, through the equation

$$F\left(s_i\right) = \frac{1}{2}\left(1 + \log_{g/2}^{(\sqrt{2})^{-1(s_i) - g/2}}\right) \tag{1}$$

where $\Delta^{-1}\left(s_i\right) = i$ $(i \in \{0, 1, \ldots, g\})$ and $g = 8$.

Definition 1 Let U be a discourse domain, and the triplet with $P = \left\{x, \mu_{p(x)}, v_{p(x)}\right\}$ on U is called a PFSs where the $\mu_{p(x)}^2 + v_{p(x)}^2 \leq 1$. When $\mu_{p(x)}^2 + v_{p(x)}^2 \geq 1$, $h = \frac{1}{2} \times (u_{p(x)} + v_{p(x)} - \sqrt{2 - \left|u_{p(x)} - v_{p(x)}\right|})$ indicates the uncertainty. Then the $u_{p(x)} = u_{p(x)} - h$ and $v_{p(x)} = v_{p(x)} - h$.

According to the above definition, and Eq. 1 for the evaluation of D_1 can be translated into the numerical matrix as follows:

$$D_1' = \begin{bmatrix} (0.5, 0.5) & (0.967830086, 0.092830086) & (0.875, 0.25) \\ (1.11E - 16, 0.875) & (0.5, 0.5) & (0.375, 0.75) \\ (0.125, 0.75) & (0.625, 0.25) & (0.5, 0.5) \end{bmatrix}$$

Second, the numerical matrix is translated to the Pythagoras Fuzzy Numbers (PFNs) by the Intuitionistic Fuzzy Archimedean t-conorm and t-norm (IFWGA) [19]. So the numerical matrix D_1' is converted to the PFNs. The evaluations of decision-makers are all converted to the PFNs as shown in Table 1.

Table 1 PFNs of alternatives for mobile phone quality assessment for decision-makers

	A_1	A_2	A_3
1	(0.42, 0.58)	(0.33, 0.67)	(0.25, 0.75)
2	(0.69, 0.50)	(0.0, 0.82)	(0.31, 0.68)
3	(0.42, 0.64)	(0.29, 0.70)	(0.28, 0.67)
4	(0.42, 0.60)	(0.18, 0.82)	(0.40, 0.58)
5	(0.36, 0.67)	(0.33, 0.67)	(0.30, 0.67)
6	(0.49, 0.61)	(0.22, 0.69)	(0.28, 0.69)
7	(0.32, 0.70)	(0.30, 0.66)	(0.37, 0.64)
8	(0.33, 0.67)	(0.25, 0.75)	(0.42, 0.58)
9	(0.39, 0.72)	(0.39, 0.71)	(0.22, 0.56)
10	(0.39, 0.61)	(0.39, 0.63)	(0.22, 0.75)

2.2 Consensus Measure

A Consensus Reaching Process (CRP) within the resolution of a GDM problem is usually proposed and widely implemented to help decision-makers achieve consensus regarding the obtained collective solution [17, 20]. Consensus measures between PFNs and the aggregated result are defined, which can measure the differences between the final solution and evaluations provided by crowd participants. In particular, the Consensus Measure Between PFNs and Aggregated Result are defined as follows:

Definition 2 Let $\alpha = (\mu_i, v_i)$ be the PFNs and $\beta = (\mu_\beta, v_\beta)$ be the aggregated result. And the Consensus Measure between Each PFN(CMEI) and the aggregated result is defined as follows:

$$CMEI(\alpha_i, \beta) = 1 - \left(\left|u_i - u_\beta\right| + \left|v_i - v_\beta\right| + \left|\pi_i - \pi_\beta\right|\right)$$

where $\pi_i = 1 - u_i - v_i, \pi_\beta = 1 - u_\beta - v_\beta$, and the π_i denotes the uncertainty of the PFNs, while the π_β denotes the uncertainty of the aggregated result.

The Consensus Measure between all PFNs (CMI) and the aggregated result is defined as follows:

$$CMI = 1 - \frac{1}{n}\sum_{i=1}^{n}\left(\left|u_i - u_\beta\right| + \left|v_i - v_\beta\right| + \left|\pi_i - \pi_\beta\right|\right)$$

3   The Proposed Method

In this chapter, the decision-makers compare the alternatives in pairwise comparison based on PFSs from the positive and negative views, and the evaluations are transformed into PFNs to avoid the problem of inconsistency. Outlier detection is used to detect the opinions of decision-makers under each alternative (which is different from the opinions of most people), and then the evaluations are aggregated using the aggregation method.

3.1   The Outlier Detection Method

For the PFNs, the outliers are deemed significantly large or small from most existing values. And the other data is called concentrate data. In [14, 21], the method of outliers detection is proposed. In case the data distribution is uniform, the results obtained by these methods are not so as good as expected. So the method of Searching the Outliers of PFNs (SOPFNs) is proposed. The SOPFNs mainly calculates the support between the member degree to divide the data into Larger Outlier Region (LOR), Smaller Outlier Region (SOR), or data Concentration Region (CR). The support becomes higher if the data is more similar. The calculation of support is as follows:

$$Support(i, j) = 1 - |i - j|$$

For example, in Table 2, The P_i denotes the decision-maker, and the u_{ij} denotes the weight of alternatives A_j given by P_i. Before aggregating the decision-makers' opinions of each alternative, the SOPFNs algorithm is performed first. For instance, the outliers of alternative A_1 need to be detected.

First, calculate the minimum value, maximum value, and average value in $\{u_{11}, u_{21}, \ldots, u_{ni}\}$. And initialize the lists support_minimun, support_maximun, support_average, which record the support between each data and the minimum value, maximum value, and average value, respectively. And initialize the lower_outliers, concentrate_data, higher_outlier which are the result of SOPFNs.

Table 2 PFNs of alternatives for decision-makers

	A_1	A_2	A_3
P_1	(u_{11},v_{11})	(u_{12},v_{12})	(u_{13},v_{13})
P_2	(u_{21},v_{21})	(u_{22},v_{22})	(u_{13},v_{13})
...	...	...	...
P_n	(u_{n1},v_{n1})	(u_{n2},v_{n2})	(u_{n3},v_{n3})

Second, calculate the support between each member degree, i.e., $\{u_{11},$ $u_{21}, \ldots, u_{ni}\}$ and minimum value. The support between the member degree and minimum is $support\_minimum = \{w_1, w_2, \ldots, w_n\}$.

Third, adjust the support between each data and the minimum value to make the difference between the support more obvious. Let $X = w_1 \times u_{11} + w_2 \times u_{21} + \cdots + w_n \times u_{n1}$ and calculate the support between the member degree and X. Then, the support_minimum' and X' is obtained. The process iterates until $\frac{X-X'}{X} < 20\%$.

Fourth, the support_minimum is obtained after the above steps. In the same steps, support_maximun and support_average can be calculated.

Fifth, compare the support of each data with the minimum, maximum, and average. If the support of u_i with the minimum is greater, it is a smaller outlier. The (u_{i1}, v_{i1}) is added to the lower_outlier. If the support of u_i with the maximum is greater, it is a larger outlier. The (u_{i1}, v_{i1}) is added to the higher_outlier. Otherwise, it is not an outlier, and the (u_{i1}, v_{i1}) is added to the concentrate_data.

Example 1 In Table 1, for the alternative A_3, the results of lower_outliers are $\{(0.25, 0.75), (0.22, 0.56), (0.22, 0.75)\}$. The results of higher_outliers are $\{(0.40, 0.58), (0.37, 0.64), (0.42, 0.58)\}$. The results of concentrate_data are $\{(0.31, 0.68), (0.30, 0.67), (0.28, 0.69)\}$.

3.2 The Aggregation Method

Many aggregation functions are proposed to aggregate the evaluations. In order to decrease the influences of outliers, the disjunctive and conjunctive functions are used to aggregate the lower outliers and larger outliers, respectively. If the value in the concentration region is in [a, b], the aggregation result by disjunctive aggregation is greater than $\max\{X_i | i = 1, 2, \ldots, n\}$ and less than a for outliers with lower values, so the impact of outliers with lower values can be reduced. For outliers with higher values, the aggregation result by conjunctive aggregation is less than $\min\{X_i | i = 1, 2, \ldots, n\}$ and greater than b for outliers with high values, so the impact of outliers with lower values can also be reduced.

Conjunctive and Disjunctive function satisfy the following conditions:

$$\lim_{x \to a, y \to a} \text{Disjunctive}(x, y) = \lim_{x \to a, y \to a} \text{Averaging}(x, y)$$
$$\lim_{x \to b, y \to b} \text{Conjunctive}(x, y) = \lim_{x \to b, y \to b} \text{Averaging}(x, y)$$

The two special classes of conjunctive and disjunctive aggregation functions are the triangular norms and conorms. Hamacher t-norms and t-conorms [22] are popular continuous Archimedean t-norm and t-conorm [23] in many practices, with well-known Algebraic and Einstein t-conorms and t-norms as the special cases [24]. For simplicity, the Hamacher t-norms and t-conorms were adopted as the special instances in this article, the following conjunctive and disjunctive functions are introduced to construct the Mixed Aggregation (MA). The following conjunctive and disjunctive functions are denoted as $MC(x, y)$ and $MD(x, y)$, which is defined as follows:

Definition 3 Based on the Hamacher t-norms and t-conorms, for $a, b \in (0, 1)$, the following can be constructed: (1) Let $MD_\gamma^H(x, y) = \dfrac{x + y - \frac{xy}{a} - (1 - \gamma)\frac{xy}{a}}{1 - (1 - \gamma)\frac{xy}{a^2}}, \gamma >$

0, $x, y \in [0, a]$, MD_γ^H is monotonously increased about γ and has the following property.

$$\lim_{x \to a, y \to a} MD_\gamma^H(x, y) = \lim_{x \to a, y \to a} \text{Averaging}(x, y)$$

(2) Let $MC_\gamma^H(x, y) = \dfrac{\frac{(x-b)(y-b)}{1-b}}{\gamma + (1-\gamma)(\frac{(x-b)}{1-b} + \frac{(y-b)}{1-b} - \frac{(x-b)(y-b)}{(1-b)^2})} + b$, $\gamma > 0$, x, y $\in [b, 1]$, MC_γ^H is

monotonously decreased about γ and has the following property.

$$\lim_{x \to b, y \to b} MC_\gamma^H(x, y) = \lim_{x \to b, y \to b} \text{Averaging}(x, y)$$

where $\text{Averaging}(x, y) = \dfrac{x + y}{2}$. And if the effects of extremely small outliers on aggregating result are more significant than the ones with extremely large outliers, for instance, the number of small outliers may be greater than the one with larger outliers or there are no large outliers, then the larger values should be chosen for γ.

And the MA is defined as follows:

Definition 4 Let $\mathbf{X} = \{x_1, x_2, \ldots, x_n\}$ denote the vector of values to be aggregated and $[a, b] \subseteq [0, 1](1 \le i \le n)$ be CR of $x_i (1 \le i \le n)$. The Multivariate MA is a mapping: $MA : [0, 1]^n \to [0, 1]$.

$$MA(X) = M(MD(x_{(1)}^l, x_{(2)}^l, \cdots, x_{(n_1)}^l), \text{Averaging}(x_{(1)}^m, x_{(2)}^m, \cdots, x_{(n_2)}^m),$$
$$EMC(x_{[1]}^h, x_{[2]}^h, \cdots, x_{[n_3]}^h)) \tag{2}$$

where $EMD(X_i) = MD(EMD(x_1, x_2, \ldots, x_{i-1}), x_i)$, $EMC(X_i) = MC(EMC$ $(x_1, x_2, \ldots, x_{i-1}), x_i)$, $(x_{(1)}^l, x_{(2)}^l, \ldots, x_{(n_1)}^l)$ is the lower outliers, and $(x_{[1]}^h, x_{[2]}^h, \ldots, x_{[n_3]}^h)$ is the large outliers.

Example 2 Depending on the SOPFNs, the result of alternative A_3 is obtained in Example 1. And the aggregation result is $(\mu_\beta, v_\beta) = (MA(u_{11}, \mu_{21}, \ldots, \mu_{n1}), MA(v_{11}, v_{21}, \ldots, v_{n1})) = (0.30, 0.68)$.

4 Experimentation and Analysis

In order to verify the effectiveness of the proposed method, this research invited 30 students to compare the evaluation indexes of mobile phone quality in a pairwise

Table 3 Aggregation results of mobile phone quality evaluation indexes

	A_1	A_2	A_3
NO.	(0.41, 0.64)	(0.28, 0.69)	(0.31, 0.67)
MMAFI	(0.38, 0.58)	(0.30, 0.72)	(0.32, 0.7)
MA	(0.42, 0.62)	(0.28, 0.7)	(0.3, 0.68)

Table 4 Consensus measure of mobile phone quality evaluation indexes

	A_1	A_2	A_3
NO.	0.8372	0.8209	0.8583
MMAFI	0.8189	0.8222	0.8428
MA	0.8393	0.8201	0.8503

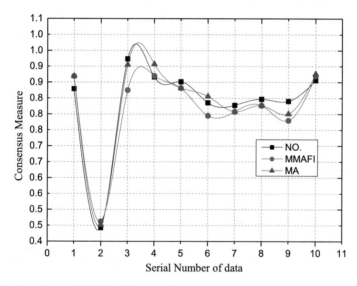

Fig. 1 Comparative results of consensus measures

manner from position and negative views. Furthermore, the evaluation results of these decision-makers are converted into PFNs, as shown in Table 1. According to the SOPFNs, outliers of expert opinions for each alternative are detected. And then the weights of the alternatives are obtained by MA.

The aggregations of students' evaluations are presented in Table 3. And the consensus measure is shown in Table 4. In Table 4, the "NO." represents the aggregation result of no outliers. In this case, the suspected outliers are all removed. The "MMAFI" is the aggregation result by the method in [14]. And the "MA" denotes the aggregation result by the proposed method. From Table 4, the consensus measure of MA is closer to the MMAFI. As shown in Fig. 1, the results of MA to alternative

A_1 are closer to the NO than MMAFI in many cases. And the orders of alternatives by the proposed and MMAFI are $A_1 \succ A_3 \succ A_2$. And in [12], the orders of alternatives are also $A_1 \succ A_3 \succ A_2$. While the results computed by the proposed method are consistent with two recent state-of-the-arts, the experimental case study demonstrates its efficacy with its effectiveness lying in the mechanism that the evaluations of decision-makers do not need to be modified or disregarded.

5 Conclusion

In order to improve individual consistency and consensus, a lot of research has been directed to modify the initial opinions from decision-makers and discard information that does not meet the consistency threshold. To solve this problem, this chapter proposes an alternative aggregation approach without modifying or discarding any original information. This is achieved by first detecting outliers, then aggregating all values using a mixed strategy, thus mitigating the issue of individual consistency and consensus. The initial case study demonstrates the efficacy and effectiveness of the proposed work. While promising, future work will explore working in combination with advanced fuzzy systems [25, 26] to further address uncertainty and imprecision from the decision-makers' linguistic evaluations.

References

1. Liao, H., Xu, Z., Zeng, X. J., & Xu, D. L. (2015). An enhanced consensus reaching process in group decision making with intuitionistic fuzzy preference relations. *Information Sciences*, p. S0020025515006787.
2. Chen, T., Su, P., Shang, C., Hill, R., Zhang, H., Shen, Q.: Sentiment classification of drug reviews using fuzzy-rough feature selection. In *2019 IEEE international conference on fuzzy systems (FUZZ-IEEE)*, pp. 1–6.
3. Chen, T., Shang, C., Su, P., Keravnou-Papailiou, E., Zhao, Y., & Antoniou, G., et al. (2020). A decision tree-initialised neuro-fuzzy approach for clinical decision support. *Artificial Intelligence in Medicine*.
4. Chen, T., Su, P., Shang, C., & Shen, Q. (2018). 'Weighted fuzzy rules optimised by particle swarm for network intrusion detection. In *2018 IEEE International Conference on Fuzzy Systems (FUZZ-IEEE)*, pp. 1–7.
5. Dong, Y., Zhao, S., Zhang, H., Chiclana, F., & Herrera-Viedma, E. (2018). A self-management mechanism for noncooperative behaviors in large-scale group consensus reaching processes. *IEEE Transactions on Fuzzy Systems*, 26(6), 3276–3288.
6. Chiclana, F., Mata, F., Martinez, L., Herreraviedma, E., & Alonso, S. (2008). Integration of a consistency control module within a consensus model. *International Journal of Uncertainty, Fuzziness and Knowledge-Based Systems*, 16, 35–53.
7. Hengshan, Z., Yimin, Z., Zheng, Q., Zhongmin, W., Yanping, C., Chunru, C., & Ting, L. (2020). A novel large group decision making method via normalized alternative prediction selection. *IEEE Transactions on Fuzzy Systems*, pp. 1–1.

8. Zhang, H., Chen, T., Wang, Z., Chen, Y., & Chen, R. (2020). A pythagorean fuzzy preference relation approach for group decision making. In *2020 IEEE international conference on fuzzy systems (FUZZ-IEEE)*.

9. Li, C.-C., Rodriguez, R. M., Martinez, L., Dong, Y., & Herrera, F. (2019). Consensus building with individual consistency control in group decision making. *IEEE Transactions on Fuzzy Systems, 27*(2), 319–332.

10. Wu, Z., & Xu, J. (2017). A consensus model for large-scale group decision making with hesitant fuzzy information and changeable clusters. *Information Fusion*, p. S1566253517301938.

11. Li, C., Dong, Y., & Herrera, F. (2019). A consensus model for large-scale linguistic group decision making with a feedback recommendation based on clustered personalized individual semantics and opposing consensus groups. *IEEE Transactions on Fuzzy Systems, 27*(2), 221–233.

12. Zhang, Z., & Pedrycz, W. (2018). Goal programming approaches to managing consistency and consensus for intuitionistic multiplicative preference relations in group decision making. *IEEE Transactions on Fuzzy Systems, 26*(6), 3261–3275.

13. Zheng, Q., Liu, T., Qu, Y., Luo, M., & Zhang, H. (2016). Improving linguistic pairwise comparison consistency via linguistic discrete regions. *IEEE Transactions on Fuzzy Systems A Publication of the IEEE Neural Networks Council*.

14. Zhang, H., Zheng, Q., Wang, Z., Chen, Y., Qu, Y., & Liu, T. (2018). Crowd intelligence for decision making based on positive and negative comparing with linguistic scale. In *2018 IEEE international conference on fuzzy systems (FUZZ-IEEE)*, pp. 1–8.

15. Zhang, H., Zheng, Q., Wang, Z., Chen, Y., Liu, T., & Chen, T. (2019). Method selecting correct one among alternatives utilizing intuitionistic fuzzy preference relation without consensus reaching process. In *2019 IEEE international conference on fuzzy systems (FUZZ-IEEE)*, pp. 1–6.

16. Chen, X., Zhang, H., & Dong, Y. (2015). The fusion process with heterogeneous preference structures in group decision making: A survey. *Information Fusion, 24*, 72–83.

17. Dong, Y., Zha, Q., Zhang, H., Kou, G., Fujita, H., Chiclana, F., & Herrera-Viedma, E. (2018). Consensus reaching in social network group decision making: Research paradigms and challenges. *Knowledge-Based Systems, 162*(15), 3–13.

18. Xu, Z. (2015). Deviation square priority method for distinct preference structures based on generalized multiplicative consistency. *IEEE Transactions on Fuzzy Systems, 23*(4), 1164–1180.

19. Zhang, H., Zheng, Q., Wang, Z., Chen, Y., & Chen, T. (2019). Method selecting correct one among alternatives utilizing intuitionistic fuzzy preference relation without consensus reaching process. In *2019 IEEE international conference on fuzzy systems (FUZZ-IEEE)*.

20. Kozierkiewicz-Hetmanska and Adrianna. (2017). The analysis of expert opinions' consensus quality. *Information Fusion, 34*, 80–86.

21. Zhang, H., Zheng, Q., Liu, T., & Cui, X. A grade assignment and ifs translation approach based on intensive region searching. In *IEEE international conference on fuzzy systems*.

22. Beliakov, G., Pradera, A., & Calvo, T. (2008). *Aggregation functions: A guide for practitioners*. Incorporated: Springer Publishing Company.

23. Nguyen, H. T., & Walker, E. A. (2005). *A first course in fuzzy logic*. CRC Press.

24. Xia, M., Xu, Z., & Zhu, B. (2012). Some issues on intuitionistic fuzzy aggregation operators based on Archimedean t-conorm and t-norm. *Knowledge-Based Systems, 31*, 78–88.

25. Chen, T., Shang, C., Yang, J., Li, F., & Shen, Q. (2019). A new approach for transformation-based fuzzy rule interpolation. *IEEE Transactions on Fuzzy Systems*, 1–1.

26. Chen, T., Shang, C., Su, P., & Shen, Q. (2018). Induction of accurate and interpretable fuzzy rules from preliminary crisp representation. *Knowledge Based Systems, 146*(15), 152–166.

Novel Aggregation Functions Based on Domain Partition with Concentrate Region of Data

Hengshan Zhang, Tianhua Chen, Zhongmin Wang, and Yanpin Chen

Abstract Combining numerous input arguments, specially in case most arguments lie in a concentrate region, is a complex issue. This chapter proposes to partition the input domain on the basis of the concentrate region, which can then be tackled based on the sub-regions. Furthermore, two bi-variate aggregation functions are proposed, which aim to behave differently in response to the corresponding sub-regions. The bi-variate functions are extended further into multivariate functions in combination with the popular Ordered Weighted Averaging *OWA* operators. Finally, the proposed aggregation functions are assessed using a case study where the maintainability of the Linux Kernels is evaluated, demonstrating the effectiveness of the proposed functions.

Keywords Aggregation function · Conjunctive · Disjunctive · Concentrate region · T-norm · T-conorm

H. Zhang (✉) · Z. Wang · Y. Chen
Shanxi Key Laboratory of Network Data Analysis and Intelligent Processing, Xi'an, China
e-mail: zhanghs@xupt.edu.cn

Z. Wang
e-mail: zmwang@xupt.edu.cn

Y. Chen
e-mail: chenyp@xupt.edu.cn

School of Computer Science and Technology, Xi'an University of Posts
and Telecommunications, Xi'an, China

T. Chen
Department of Computer Science, School of Computing and Engineering, University
of Huddersfield, Huddersfield, UK
e-mail: T.Chen@hud.ac.uk

© Springer Nature Switzerland AG 2021
J. Carter et al. (eds.), *Fuzzy Logic*,
https://doi.org/10.1007/978-3-030-66474-9_8

1 Introduction

Aggregating information from different sources into a single numerical value is an important and complex issue. It has received extensive attention from researchers and practitioners over the past decades. And a variety of functions have been developed to aggregate information under various environments [1–7].

There are many situations in practice, where the majority of data is distributed in a definite region [a, b]. The data scattered outside of [a, b] may be the noises or special cases. For example, the "Halstead size" is a metric measuring software quality. We obtained the values by analyzing 880 Linux Kernels and noticed that over 80% values locate in the concentrate region (*CR*) [0.32, 0.62]. The values which are not in the concentrate region are generally very small or large. As an example, we randomly choose some values (represent as set *R*) from 880 data. $R = \{0.92, 0.9054, 0.6088, 0.5963, 0.5881, 0.5898, 0.5965, 0.5978, 0.5872, 0.5797, 0.0357, 0.0002\}$.

The values {0.92, 09054} exceed the higher boundary of concentrate region, and {0.0357, 0.0002} are smaller than the lower boundary. It is not straightforward to judge whether these values are special cases or noisy data. As a result, we aim to decrease the effects of the values that are not in *CR* and to get a more reasonable combined result. Alternatively, we could make use of these values based on their influence over the aggregated result.

In the literature, the four classes of aggregation operators are averaging, conjunctive, disjunctive, and mixed aggregation operators [1]. The use of the averaging, conjunctive, and disjunctive operators may not distinguish the data located in or outside *CR*. Mixed aggregation operators are those whose behavior depends on the input values. These functions exhibit conjunctive, disjunctive, or averaging behavior on different parts of their domain [1]. *MYCIN* [8] is a well-known expert system, in which the inputs that are smaller than 1/2 represent the "negative" evidences, and larger than 1/2 representing the "positive" evidences. Scuh aggregation function exhibits conjunctive behavior on $[0, 1/2]^2$ and disjunctive behavior on $[1/2, 1]^2$. For the rest of domain, the aggregation is simple averaging.

The mixed aggregation functions include various families [1], including well-known ones such as the uninorms [9, 10], nullnorms [11], and *ST-OWAs* [12, 13] that are related to triangular norms and conforms [14, 15]. In particular, the uninorms are associative aggregation functions which make conjunctive operator when dealing with lower input values, disjunctive operator for higher values and averaging otherwise [1]. On the other hand, the nullnorms are the disjunctive operator for lower values, and conjunctive operator for higher values [1]. The operators of uninorms and nullnorms could be shown as Fig. 1 [1].

In this chapter, we consider the situation where two attributes are combined by using the mixed aggregation functions. The most values of the attributes are in the concentrate region *CR*. The domain of functions needs to be partitioned into some sub-regions by using *CR* in this situation. A possibility partition may be shown as Fig. 2a.

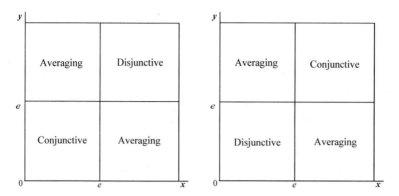

Fig. 1 Behavior of uninorms (left) and nullnorms (right)

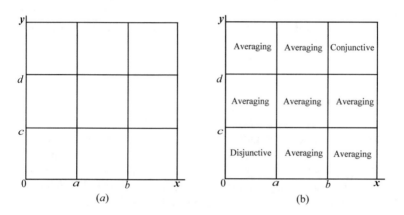

Fig. 2 Partition of the domain (**a**) and behavior of an aggregation function (**b**)

To this problem, the previously reviewed mixed aggregation functions may not work well. Instead we propose to partition the domain of function based on the distribution using CR of the data and classify the partitioned sub-regions of the domain. We therefore propose two binary aggregation functions, which have different behaviors in response to different class sub-regions. One binary aggregation function aims to reduce the effects of data outside of CR, with the potential behavior shown in Fig. 2b. Whereas the other binary function aims to increase the effects of these data. Furthermore, we extend them into multivariate functions, resulting in four aggregation functions, which allows to combine with the ordered weighted averaging OWA and other aggregation functions. Assuming the weight of every value in R is 0.0833, and the conjunctive and disjunctive functions are Einstein t-norm (represent as ET) and t-conorm (represent as ES). We can then aggregate the values in R into a single number as

$$ET(0.92, 0.9054) + OWA(0.6088, 0.5963, 0.5881, 0.5898,$$
$$0.5965, 0.5978, 0.5872, 0.5797) + ES(0.0357, 0.0002) = 0.5457.$$

The main contributions of this chapter include (1) we partition the domain of function into the sub-regions based on the distribution using CR of the data, and classify the partitioned sub-regions. (2) We proposed two bi-variate aggregation functions, which have different behaviors in response to different class sub-regions. They can decrease or increase the effects of the data outside of CR. (3) We extend the bi-variate aggregation functions into multivariate functions and combine with the Ordered Weighted Averages (*OWA*), which leads to novel useful *OWA* aggregation functions. (4) We evaluate the maintainability of the Linux Kernels using generated *OWA* aggregation functions in an experimental study.

The rest of this chapter is organized as follows. Section II briefly reviews related works. We represent the proposed aggregation functions in Section III. The evaluation of the maintainability for Linux Kernels is conducted in section IV. Section V concludes the chapter and points out future work.

2   Related Works

The main categories of aggregation function include averaging, conjunctive, disjunctive, and mixed functions [1]. Commonly used operators such as the arithmetic mean, weight arithmetic mean, geometric mean, harmonic mean and medians, are typically used averaging aggregation functions [1]. The Ordered Weighted Averaging (OWA) is a parameterised operator based on the ordering of extraneous variables to which it is applied and has been successfully applied in various domains [16, 17]. Recent developments have advanced the proposals of numerous variants of OWA-based aggregation functions. Well-known examples include the ordered-weighted-geometric function [18, 19], weighted *OWA* function[20], induced *OWA* function [21], induced ordered weighted-geometric function [2], uncertain *OWA* function [22], hybrid aggregation function [23], fuzzy-weighted-average function [24], generalized *OWA* aggregation function [25].

In particular, Yager [26] developed a power average (*PA*) function and a power *OWA (POWA)* function, whose weighting vectors depend upon the input arguments and allow values being aggregated to support and reinforce each other. On top of that, Z.Xu [4] proposed a power geometric (*PG*) function based on *PA* function and the geometric mean, and a power-ordered-geometric (*POG*) function and a power-ordered weighted-geometric (*POWG*) function based on power *OWA (POWA)* function and the geometric mean.

Yager [5] proposed a prioritized aggregation operator (*PAO*) to handle the situation where lack of satisfaction to criteria denoted as higher priority cannot be compensated by increased satisfaction by those denoted as lower priority. Yager [27] also introduced a probabilistically weighted *OWA* aggregation to obtain a representative value through aggregating the information in the uncertain profile. This value is used to select one from a collection of alternatives. Two special and well-known classes of conjunctive and disjunctive aggregation functions are the triangular norms and conorms [1, 14, 15, 28, 29]. Archimedean t-conorm and t-norm are generalizations of commonly used t-conorms and t-norms, such as algebraic, Einstein, with the continuous Archimedean t-norms and t-norms being very popular, especially in intuitionistic fuzzy environment [30].

In some applications, the way an aggregation functions behave is subject to the input arguments, where the above-reviewed methods generally do not adapt well. To overcome this issue, the mixed aggregation functions are proposed and studied [1] in the literature. Uninorms [9, 10], nullnorms [11], and ST-OWAs [12, 13] are some important mixed aggregation functions. ST-OWAs [31–33] and U-OWAs [1] are two mixed aggregations on the basis of OWA [18] operators. In [6], the authors propose an aggregation function, in which at least one input must exceed a threshold in order to achieve a nonzero aggregation output; otherwise the aggregation output takes its maximum value, if any one of the inputs exceeds a corresponding threshold. However, these mixed aggregation functions do not well work in the situation where the inputs are partitioned by CR.

3 Proposed Mixed Aggregation Functions

3.1 Analyzing and Classifying Partitioned Sub-regions of Domain

The definition of CR may include three cases based on the relative relationship between CR and the whole data domain. In general, these cases are shown in Fig. 3. Specifically, in Fig. 3, [a, b] is CR, and [u, v] is the whole region of the data. In Fig. 3a, the end points of CR do not overlap with the end points of the whole region. In Fig. 3b, the point a of CR overlaps with the point u. In Fig. 3c, the point b of CR overlaps with the point v. We consider the bi-variate situation. We analyze and classify the partitions of domain. There are four cases in this situation. These cases are shown in Fig. 4a–i. The axis x and axis y represent two attributes whose values need to be combined into a single number in Fig. 4.

The partition of the domain can include six cases in Fig. 5. The case 1 is that two values are below CR. The case 2 is that a value is below CR, but another value is in CR. The case 3 is that two values are in CR. The case 4 is that a value is in CR, but

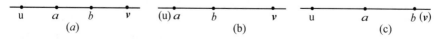

Fig. 3 Distributions for the concentrate region of the data

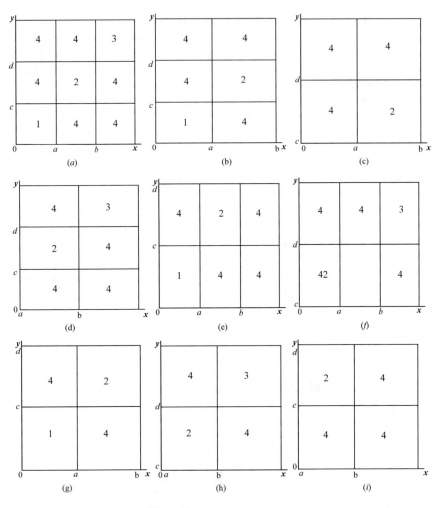

Fig. 4 Classification for the partition of domain

another value is upon *CR*. The case 5 is that two values are upon *CR*. The case 6 is that a value is below *CR*, but another value is upon *CR*. The case 1 and case 2 can be combined into one. We name this combined case as situation I. A value is below *CR*, but another value is in or below *CR* situation I. We name case 3 as situation II. The case 4 and case 5 can be combined into one. In this combined case, a value is upon *CR*, another value is in or upon *CR*. We named this combined case as situation III. In case 6, we suggest two strategies to apply the aggregation functions. The first one is to apply the averaging function, and another is to apply the disjunctive function or the conjunctive function according to the priority of the attributes. If the attribute with the larger value has the higher priority, then the users apply the disjunctive function, while the influence of the attribute is increased, or apply the conjunctive function,

Fig. 5 Cases for the partition of domain

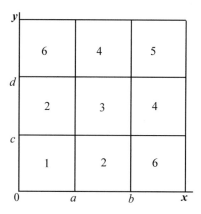

Table 1 Formulations of the classified situations

Situation	Description	Formulation of situation
I	One value is below CR, another value is in CR or below CR	$\min(x, y) < a, \max(x, y) < b$
II	Two values are in CR	$a \leq \min(x, y) \leq \max(x, y) \leq b$
III	One value is upon CR, another value are in CR or upon CR	$\min(x, y) > a, \max(x, y) > b$
IV	One value is upon CR, another value is below CR	$\min(x, y) < a, \max(x, y) > b$

while the influence of the attribute is decreased. If the attribute with the smaller value has the higher priority, the conditions are contrary. We name this case as situation IV, where we apply the averaging function. We summarize the classifications to the partitioned sub-regions of the domain for all cases and represent them in Fig. 4.

CRs of the aggregated values are different in practice. In order to describe the proposed new aggregation functions, we suppose CRs of the values are same one. If CR of the aggregated values is the region [a, b], then we can summarize and formulate these classified situations as Table 1.

3.2 Bi-variate Aggregation Functions

In this section, we consider the definitions of two bi-variate aggregation functions. The range of the aggregated values can be transformed into [0, 1], assuming unit interval is used.

The purpose of *DEGF* (Decreasing Effects Aggregation Function) is to decrease the effects of the values that are outside CR. The behavior of *DEGF* is shown in Fig. 6a.

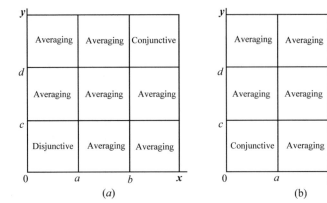

Fig. 6 Behaviors of DEGF (**a**) and IEGF (**b**)

Definition 3.1 (DEGF) Let x and y denote the values to aggregate; x and y have a concentrate region (CR) [a, b]. The Decreasing effects aggregation function (DEGF) is a mapping, i.e., $DE : [0, 1]^2 \rightarrow [0, 1]$, defined as

$$DE(x, y) = \begin{cases} \text{conjunctive} & \max(x, y) > b, \min(x, y) > a \\ \text{disjunctive} & \min(x, y) < a, \max(x, y) < b \\ \text{averaging} & \text{otherwise.} \end{cases} \quad (1)$$

The purpose of *IEGF* (Increasing Effects Aggregation Function) is to increase the effects of the values that are outside *CR*. It is the dual function of *DEGF*. The behavior of *IEGF* is shown in Fig. 6b.

Definition 3.2 (IEGF) Let x and y denote the values to aggregate; x and y have a concentrate region (CR) [a, b]. An Increasing effect aggregation function is a mapping, i.e., $IE : [0, 1]^2 \rightarrow [0, 1]$, defined as

$$IE(x, y) = \begin{cases} \text{disjunctive} & \max(x, y) > b, \min(x, y) > a \\ \text{conjunctive} & \min(x, y) < a, \max(x, y) < b \\ \text{averaging} & \text{otherwise.} \end{cases} \quad (2)$$

The continuous Archimedean t-norms and t-conorms [34] are very useful. The main reasons are that (a) they form a dense subset in the set of all continuous t-norms or t-comorms, and (b) they can be represented via additive and multiplicative generators [34]. Einstein t-norm and t-conorm are the continuous Archimedean t-norm and t-conorm. If applying them, then we can get the definite and useful *DEGF* and *IEGF*. They are the following examples.

Example 1 Let $[a, b] \subseteq [0, 1]$ is the concentrate region, the conjunctive and disjunctive functions are the Einstein t-norm (represent as ET) and t-conorm (represent as ES), the averaging function is the arithmetic mean. A useful $DEGF$ (This is named as $Ein\text{-}DEGF$) is represented as

$$Ein\_DE(x, y) = \begin{cases} \frac{xy}{1+(1-x)(1-y)} & \max(x, y) > b, \min(x, y) > a \\ \frac{x+y}{1+xy} & \min(x, y) < a, \max(x, y) < b \\ \frac{x+y}{2} & \text{otherwise.} \end{cases} \quad (3)$$

$Ein\text{-}DEGF$ has the following properties:

Proposition 3.1 $Ein\text{-}DEGF$ is symmetry: $Ein\_DE(x, y) = Ein\_DE(y, x)$.

Example 2 If the conditions are same as the example 1, an IEGF (Ein-IEGF) is the following:

$$Ein\_IE(x, y) = \begin{cases} \frac{x+y}{1+xy} & \max(x, y) > b, \min(x, y) > a \\ \frac{xy}{1+(1-x)(1-y)} & \min(x, y) < a, \max(x, y) < b \\ \frac{x+y}{2} & \text{otherwise.} \end{cases} \quad (4)$$

This function has the following properties:

Proposition 3.2 $Ein\text{-}IEGF$ is symmetry: $Ein\_IE(x, y) = Ein\_IE(y, x)$.

Proposition 3.3 Let $X = (x_1, y_1) \in [0, 1]^2$, $Y = (x_2, y_2) \in [0, 1]^2$. If $X \leq Y, X \neq Y$, then $Ein\_IE(X) \leq Ein\_IE(Y)$. That is to say, the Ein-IEGF is monotonic increasing.

Proof Without losing generality, we assume that $x_1 \leq x_2, y_1 < y_2$. The Einstein t-norm and t-conorm, and the averaging function are monotonicity in their domains. As a result, we need to proof that $Ein\text{-}IEGF$ is monotonicity while the (x_1, y_1) and (x_2, y_2) are in the different sub-domains. There are three cases.

Case 1: If $a \leq \min(x_1, y_1) \leq \max(x_1, y_1) \leq b$ (situation II), or $\max(x_1, y_1) > b$, $\max(x_1, y_1) < a$ (situation IV), then (x_2, y_2) must satisfy $\max(x_2, y_2) > b$, $\min(x_2, y_2) > b$ (situation III). As a result, $Ein\_IE(X) = \frac{x_1+y_1}{2}$, and $Ein\_IE(Y) = ES(Y)$.

$$Ein\_IE(X) = \frac{x_1+y_1}{2} \leq \frac{x_2+y_2}{2} \leq \frac{x_2+y_2}{1+x_2y_2} = Ein\_IE(Y).$$

Case 2: If $\max(x_1, y_1) < b, \min(x_1, y_1) < b$ (situation I), (x_2, y_2) may satisfy $a \leq \min(x_2, y_2) \leq \max(x_2, y_2) \leq b$ (situation II), or $\max(x_2, y_2) > b, \min(x_2, y_2) < a$ (situation IV). As a result, $Ein\_IE(X) = ET(X)$, and $Ein\_IE(Y) = \frac{x_2+y_2}{2}$.

$$Ein\_IE(X) = \frac{x_1y_1}{1+(1-x_1)(1-y_1)} \leq \frac{x_1+y_1}{2} \leq \frac{x_2+y_2}{2} = Ein\_IE(Y).$$

Case 3: If $\max(x_1, y_1) < b$, $\min(x_1, y_1) < b$ (situation I), (x_2, y_2) may satisfy $\max(x_2, y_2) > b$, $\min(x_2, y_2) > b$ (situation III). As a result, $Ein\_IE(X) = ET(X)$, and $Ein\_IE(Y) = ES(X)$.

$$Ein\_IE(X) = \frac{x_1 y_1}{1 + (1 - x_1)(1 - y_1)} \leq \frac{x_1 + y_1}{2} \leq \frac{x_2 + y_2}{2} \leq \frac{x_2 + y_2}{1 + x_2 y_2} = Ein\_IE(Y).$$

In summary, we get the *Ein-IEGF* is monotonicity.   □

3.3 Multivariate Aggregation Functions

We consider the case that the aggregated values are the collection $X = (x_1, x_2, \ldots, x_n)$, and any $x_i (1 \leq i \leq n)$ has a concentrate region $[a_i, b_i](1 \leq i \leq n)$. To this case, we give some new mixed aggregation functions which are the extension of DEGF and IEGF. Furthermore, we combine OWA operator with these aggregation functions to get some useful aggregation functions.

3.3.1 Multivariate Aggregation Functions and Properties

In the following, we represent the symbolic and their meanings used in this section.

(1). Let the symbolic (i) represents the index of the ith largest value of $X = (x_1, x_2, \ldots, x_n)$.

(2). $x_{(i)}^l < a_{(i)} (0 \leq (i) \leq n_1)$, where $x_{(i)}^l$ is the ith largest of the values which satisfy $x_i < a_i$, and n_1 is the number of the values $(x_i < a_i)$.

(3). $a_{(j)} \leq x_{(j)}^m \leq b_{(j)} (0 \leq (j) \leq n_2)$, where $x_{(j)}^m$ is the jth largest of the values which satisfy $a_i \leq x_i \leq b_i$, and n_2 is the number of the values $(a_i \leq x_i \leq b_i)$.

(4). $x_{(k)}^h > b_{(k)} (0 \leq (k) \leq n_3)$, where $x_{(k)}^h$ is the kth largest of the values which satisfy $x_i > b_i$, and n_3 is the number of the values $(x_i > b_i)$. In addition, $n_1 + n_2 + n_3 = n$.

Definition 3.3 (E-DEGF). Let $X = (x_1, x_2, \ldots, x_n)$, and $[a_i, b_i](1 \leq i \leq n)$ is the concentrate region of $x_i (1 \leq i \leq n)$. An extended *DEGF (E-DEGF)* is a mapping: $EDE : [0, 1]^n \to [0, 1]$.

$$EDE(X) = \text{Averaging}(\text{Disjunctive}(x_{(1)}^l, x_{(2)}^l, \ldots, x_{(n_1)}^l),$$
$$\text{Averaging}(x_{(1)}^m, x_{(2)}^m, \ldots, x_{(n_2)}^m), \text{Conjunctive}(x_{(1)}^h, x_{(2)}^h, \ldots, x_{(n_3)}^h)). \quad (5)$$

Definition 3.4 (E-IEGF). Let $X = (x_1, x_2, \ldots, x_n)$, and $[a_i, b_i](1 \leq i \leq n)$ is the concentrate region of $x_i (1 \leq i \leq n)$. An extended *IEGF (E-IEGF)* is a mapping: $EIE : [0, 1]^n \to [0, 1]$.

$$EIE(X) = \text{Averaging}(\text{Conjunctive}(x^l_{(1)}, x^l_{(2)}, \ldots, x^l_{(n_1)}),$$
$$\text{Averaging}(x^m_{(1)}, x^m_{(2)}, \ldots, x^m_{(n_2)}), \text{Disjunctive}(x^h_{(1)}, x^h_{(2)}, \ldots, x^h_{(n_3)})). \quad (6)$$

In practice, the correlation between the experimental data and results may be the positive correlation. So the users wish to know the enhanced effects of the values which are below the concentrate region. On the contrary, while the correlation is negative, the users wish to improve the influence of the values that are upon the concentrate region. As a result, we give the following two aggregation functions to depict these cases. We named them as Enhanced the effects of Below Concentrate Region data (*EBCR*) and Enhanced the effects of Upon Concentrate Region data (*EUCR*).

Definition 3.5 (EBCR) Let $X = (x_1, x_2, \ldots, x_n)$, and $[a_i, b_i](1 \le i \le n)$ is the concentrate region of $x_i(1 \le i \le n)$. The definition of the aggregation function that enhanced the effects of the values that are below the concentrate regions is following:

$$EBCR(X) = \text{Averaging}(\text{Conjunctive}(x^l_{(1)}, x^l_{(2)}, \ldots, x^l_{(n_1)}),$$
$$\text{Averaging}(x^m_{(1)}, x^m_{(2)}, \ldots, x^m_{(n_2)}), \text{Conjunctive}(x^h_{(n_3)}, x^h_{(2)}, \ldots, x^h_{(1)})). \quad (7)$$

Definition 3.6 (EUCR) Let $X = (x_1, x_2, \ldots, x_n)$, and $[a_i, b_i](1 \le i \le n)$ is the concentrate region of $x_i(1 \le i \le n)$. The definition of the aggregation function that enhanced the effects of the values that are upon the concentrate regions is following:

$$EUCR(X) = \text{Averaging}(\text{Disjunctive}(x^l_{(1)}, x^l_{(2)}, \ldots, x^l_{(n_1)}),$$
$$\text{Averaging}(x^m_{(1)}, x^m_{(2)}, \ldots, x^m_{(n_2)}), \text{Disjunctive}(x^h_{(n_3)}, x^h_{(2)}, \ldots, x^h_{(1)})). \quad (8)$$

Example 3 In case the Conjunctive function is the Einstein t-norm (*ET*), the Disjunctive function is the Einstein t-conorm (*ES*), and the Averaging function is arithmetic mean. An extended *DEGF* can be obtained as the following:

$$EDE_{ES,ET,A}(X) = \frac{1}{3}\left(ET(x^l_{(1)}, x^l_{(2)}, \ldots, x^l_{(n_1)}) + \left(\frac{x^m_{(1)} + x^m_{(2)} + \ldots + x^m_{(n_2)}}{n_2}\right)\right.$$
$$\left. + \left(ESx^h_{(1)}, x^h_{(2)}, \ldots, x^h_{(n_3)}\right)\right). \quad (9)$$

Similarly, an extended *IEGF* is as the following:

$$EIE_{ES,ET,A}(X) = \frac{1}{3}\left(ET(x^l_{(1)}, x^l_{(2)}, \ldots, x^l_{(n_1)}) + \left(\frac{x^m_{(1)} + x^m_{(2)} + \ldots + x^m_{(n_2)}}{n_2}\right)\right.$$
$$\left. + \left(ES(x^h_{(1)}, x^h_{(2)}, \ldots, x^h_{(n_3)})\right)\right). \quad (10)$$

A useful *EBCR* and *EUCR* are as follows:

$$EBCR_{ES,ET,A}(X) = \frac{1}{3}\left(ET(x^l_{(1)}, x^l_{(2)}, \ldots, x^l_{(n_1)}) + \left(\frac{x^m_{(1)} + x^m_{(2)} + \ldots + x^m_{(n_2)}}{n_2}\right)\right.$$
$$\left. + \left(ET(x^h_{(1)}, x^h_{(2)}, \ldots, x^h_{(n_3)})\right)\right). \tag{11}$$

$$EUCR_{ES,ET,A}(X) = \frac{1}{3}\left(ES(x^l_{(1)}, x^l_{(2)}, \ldots, x^l_{(n_1)}) + \left(\frac{x^m_{(1)} + x^m_{(2)} + \ldots + x^m_{(n_2)}}{n_2}\right)\right.$$
$$\left. + \left(ES(x^h_{(1)}, x^h_{(2)}, \ldots, x^h_{(n_3)})\right)\right). \tag{12}$$

These functions have the following properties:

Proposition 3.4 $EIE_{ES,ET,A}(X)$ *is monotonic increasing.*

Proof Let $X = (x_1, x_2, \ldots, x_n), Y = (y_1, y_2, \ldots, y_n)$, and $x_i \le y_i (i = 1, 2, \ldots, n)$. We have

$$EIE_{ES,ET,A}(X) = \frac{1}{3}\left(ET(x^l_{(1)}, x^l_{(2)}, \ldots, x^l_{(n_1)}) + \left(\frac{x^m_{(1)} + x^m_{(2)} + \cdots + x^m_{(n_2)}}{n_2}\right)\right.$$
$$\left. + \left(ES(x^h_{(1)}, x^h_{(2)}, \ldots, x^h_{(n_3)})\right)\right)$$
$$\le \frac{1}{3}\left(ET\left(y^l_{(1)}, y^l_{(2)}, \ldots, x^l_{(n_1)}\right) + \left(\frac{y^m_{(1)} + y^m_{(2)} + \cdots + y^m_{(n_2)}}{n_2}\right)\right.$$
$$+ ES\left(y^h_{(1)}, y^h_{(2)}, \ldots, y^h_{(n_3)}\right)\right) \le EIE_{ES,ET,A}(X). \qquad \square$$

Example 4 The "Modularity" is an important sub-feature of the software maintainability in the *ISO/IEC* 25010 software quality model. Its metrics and the about information are shown in the Table 2. The low and high bounder of the metrics is got by analyzing the 880 Linux Kernels. The analyzing method is represented in the experiment section. The values of the metrics corresponding to the Linux Kernel version V.1.0 are translated into the numbers in the unit interval [0, 1]. These numbers are the values in Table 2.

Based on Table 2, we can get that the values 0.0473 and 0.4068 are below the concentrate region, and the values 0.8741, 0.9756, 0.9817 are upon the concentrate region. The other values are in the concentrate region. Let the t-norm and t-conorm are the Einstein t-norm and t-conorm, and the *ET* and *ES* denote them, respectively. The process of the *E-DEGF* can be represented as following:

$$EDE(X) = \frac{1}{3}(ES(0.4068, 0.0473) + \frac{1}{6}(0 + 1.0 + 1.0 + 1.0 + 1.0 + 1.0)$$
$$+ ET(0.8741, 0.9756, 0.9817)) = \frac{1}{3}(0.4456 + 0.8333 + 0.8323) = 0.7037.$$

Table 2 Information of the metric included in "Modularity" for Linux Kernel V.1.0

Metrics	Values	Low bounder	High bounder
Halstead size	0.9756	0.32	0.61
Halstead volume	0.9817	0.28	0.61
Halstead difficulty	0.0473	0.32	1.0
Average cyclomatic	0	0	1.0
Essential complexity	1.0	0	1.0
Max nesting level	1.0	0	1.0
Count line code exe	1.0	0.04	1.0
Network size	1.0	0	1.0
Number network edges	1.0	0	1.0
Clustering coefficient	0.8741	0	0.15
Largest sub-graph nodes	0.4068	0.86	0.93

The process of the *E-IEGF* is

$$\text{EIE}(X) = \frac{1}{3}(\text{ET}(0.4068, 0.0473) + \frac{1}{6}(0 + 1.0 + 1.0 + 1.0 + 1.0 + 1.0)$$
$$+ \text{ES}(0.8741, 0.9756, 0.9817)) = \frac{1}{3}(0.0123 + 0.8333 + 1.0) = 0.6152.$$

The process of the *EBCR* is

$$\text{EBCR}(X) = \frac{1}{3}(\text{ET}(0.4068, 0.0473) + \frac{1}{6}(0 + 1.0 + 1.0 + 1.0 + 1.0 + 1.0)$$
$$+ \text{ET}(0.8741, 0.9756, 0.9817)) = \frac{1}{3}(0.0123 + 0.8333 + 0.8323) = 0.5593.$$

The process of the *EUCR* is

$$\text{EUCR}(X) = \frac{1}{3}(\text{ES}(0.4068, 0.0473) + \frac{1}{6}(0 + 1.0 + 1.0 + 1.0 + 1.0 + 1.0)$$
$$+ \text{ES}(0.8741, 0.9756, 0.9817)) = \frac{1}{3}(0.4456 + 0.8333 + 1.0) = 0.7596.$$

3.3.2 Combined OWA with Multivariate Aggregation Functions

In this sub-section, we combine the *OWA* operator with the above mentioned multivariate aggregation functions to get some useful functions.

Definition 3.7 (*OWA-DEGF*) Let $X = (x_1, x_2, \ldots, x_n)$, and $[a_i, b_i](1 \le i \le n)$ is the concentrate region of $x_i(1 \le i \le n)$. $W = (w_1, w_2, \ldots, w_n)$ s a weight vector, and $w_i(1 \le i \le n)$ is the weight of $x_i(1 \le i \le n)$. The aggregation function $ODE_{W,E-DEGF} : [0, 1]^n \to [0, 1]$ is defined as

$$\text{ODE}_{W,E-\text{DEGF}}(X) = \sum_{(i)=1}^{n_1} w_{(i)} \text{Disjunctive}(x_{(1)}^l, x_{(2)}^l, \ldots, x_{(i)}^l)$$

$$+ \sum_{(j)=1}^{n_2} w_{(j)} x_{(j)} + \sum_{(k)=1}^{n_3} w_{(k)} \text{Conjunctive}(x_{(n_3)}^h, x_{(n_3-1)}^h, \ldots, x_{(n_3-k+1)}^h). \quad (13)$$

Definition 3.8 (*OWA-IEGF*) Let $X = (x_1, x_2, \ldots, x_n)$, and $[a_i, b_i](1 \le i \le n)$ is the concentrate region of $x_i(1 \le i \le n)$. $W = (w_1, w_2, \ldots, w_n)$ is a weight vector, and $w_i(1 \le i \le n)$ is the weight of $x_i(1 \le i \le n)$. The aggregation function $OIE_{W,E-DEGF} : [0, 1]^n \to [0, 1]$ is defined as

$$\text{OIE}_{W,E-\text{IEGF}}(X) = \sum_{(i)=1}^{n_1} w_{(i)} \text{Conjunctive}(x_{(1)}^l, x_{(2)}^l, \ldots, x_{(i)}^l) + \sum_{(j)=1}^{n_2} w_{(j)} x_{(j)}$$

$$+ \sum_{(k)=1}^{n_3} w_{(k)} \text{Disjunctive}(x_{(n_3)}^h, x_{(n_3-1)}^h, \ldots, x_{(n_3-k+1)}^h). \quad (14)$$

Definition 3.9 (*OWA-EBCR*) Let $X = (x_1, x_2, \ldots, x_n)$, and $[a_i, b_i](1 \le i \le n)$ is the concentrate region of $x_i(1 \le i \le n)$. $W = (w_1, w_2, \ldots, w_n)$ is a weight vector, and $w_i(1 \le i \le n)$ is the weight of $x_i(1 \le i \le n)$. The aggregation function $OEBCR_{W,EBCR} : [0, 1]^n \to [0, 1]$ is defined as

$$\text{OEBCR}_{W,\text{EBCR}}(X) = \sum_{(i)=1}^{n_1} w_{(i)} \text{Conjunctive}(x_{(1)}^l, x_{(2)}^l, \ldots, x_{(i)}^l) + \sum_{(j)=1}^{n_2} w_{(j)} x_{(j)}$$

$$+ \sum_{(k)=1}^{n_3} w_{(k)} \text{Conjunctive}\left(x_{(n_3)}^h, x_{(n_3-1)}^h, \ldots, x_{(n_3-k+1)}^h\right). \quad (15)$$

Definition 3.10 (*OWA-EUCR*) Let $X = (x_1, x_2, ldots, x_n)$, and $[a_i, b_i](1 \le i \le n)$ is the concentrate region of $x_i(1 \le i \le n)$. $W = (w_1, w_2, \ldots, w_n)$ is a weight vector, and $w_i(1 \le i \le n)$ is the weight of $x_i(1 \le i \le n)$. The aggregation function $OEUCR_{W,EUCR} : [0, 1]^n \to [0, 1]$ is defined as

$$\text{OEUCR}_{\text{W,EUCR}}(X) = \sum_{(i)=1}^{n_1} w_{(i)} \text{Disjunctive}(x_{(1)}^l, x_{(2)}^l, \ldots, x_{(i)}^l) + \sum_{(j)=1}^{n_2} w_{(j)} x_{(j)}$$

$$+ \sum_{(k)=1}^{n_3} w_{(k)} \text{Disjunctive}\left(x_{(n_3)}^h, x_{(n_3-1)}^h, \ldots, x_{(n_3-k+1)}^h\right). \tag{16}$$

Example 5 If the conjunctive function is the Einstein t-norm, and the disjunctive function is the Einstein t-conorm, and the averaging function is arithmetic mean. A definite *OWA-DEGF* can be represented as the following:

$$\text{ODE}_{\text{W,E-DEGF}}(X) = \sum_{(i)=1}^{n_1} w_{(i)} \text{ET}(x_{(1)}^l, x_{(2)}^l, \ldots, x_{(i)}^l) + \sum_{(j)=1}^{n_2} w_{(j)} x_{(j)}$$

$$+ \sum_{(k)=1}^{n_3} w_{(k)} \text{ES}(x_{(n_3)}^h, x_{(n_3-1)}^h, \ldots, x_{(n_3-k+1)}^h). \tag{17}$$

Similarly, a definite *OWA-DEGF* can be represented as following:

$$\text{OIE}_{\text{W,E-IEGF}}(X) = \sum_{(i)=1}^{n_1} w_{(i)} \text{ET}(x_{(1)}^l, x_{(2)}^l, \ldots, x_{(i)}^l) + \sum_{(j)=1}^{n_2} w_{(j)} x_{(j)}$$

$$+ \sum_{(k)=1}^{n_3} w_{(k)} \text{ES}(x_{(n_3)}^h, x_{(n_3-1)}^h, \ldots, x_{(n_3-k+1)}^h). \tag{18}$$

Two useful *OWA-EBCR* and *OWA-EUCR* can be represented as following:

$$\text{OEBCR}_{\text{W,EBCR}}(X) = \sum_{(i)=1}^{n_1} w_{(i)} \text{ET}(x_{(1)}^l, x_{(2)}^l, \ldots, x_{(i)}^l) + \sum_{(j)=1}^{n_2} w_{(j)} x_{(j)}$$

$$+ \sum_{(k)=1}^{n_3} w_{(k)} \text{ET}(x_{(n_3)}^h, x_{(n_3-1)}^h, \ldots, x_{(n_3-k+1)}^h). \tag{19}$$

$$\text{OEUCR}_{\text{W,EUCR}}(X) = \sum_{(i)=1}^{n_1} w_{(i)} \text{ES}(x_{(1)}^l, x_{(2)}^l, \ldots, x_{(i)}^l) + \sum_{(j)=1}^{n_2} w_{(j)} x_{(j)}$$

$$+ \sum_{(k)=1}^{n_3} w_{(k)} \text{ES}(x_{(n_3)}^h, x_{(n_3-1)}^h, \ldots, x_{(n_3-k+1)}^h). \tag{20}$$

Example 6 The information about the "Modularity" is the same as the Example 4. We assumption the weight of every metric is 0.0909. The process of the *OWA-DEGF* can be represented as follow:

$$\begin{aligned} \text{ODE}_{\text{W,E-DEGF}}(X) &= 0.4068 \times 0.0909 + \text{ES}\,(0.4068, 0.0473) \\ &\quad \times 0.0909 + 5.0 \times 0.0909 + 0.8741 \times 0.0909 + \text{ET}(0.8741, 0.9756) \\ &\quad \times 0.0909 + \text{ET}(0.8741, 0.9756, 0.9817) \times 0.0909 \\ &= 0.037 + 0.0405 + 0.4545 + 0.0795 + 0.0773 + 0.0757 = 0.7644. \end{aligned}$$

The process of the *OWA-IEGF* is

$$\begin{aligned} \text{OIE}_{\text{W,E-IEGF}}(X) &= 0.4068 \times 0.0909 + \text{ET}\,(0.4068, 0.0473) \\ &\quad \times 0.0909 + 5.0 \times 0.0909 + 0.8741 \times 0.0909 + \text{ES}(0.8741, 0.9756) \\ &\quad \times 0.0909 + \text{ES}(0.8741, 0.9756, 0.9817) \times 0.0909 \\ &= 0.037 + 0.0011 + 0.4545 + 0.0795 + 0.0908 + 0.0909 = 0.7538. \end{aligned}$$

The process of the *OWA-EBCR* is

$$\begin{aligned} \text{OEBCR}_{\text{W,EBCR}}(X) &= 0.4068 \times 0.0909 + \text{ET}\,(0.4068, 0.0473) \\ &\quad \times 0.0909 + 5.0 \times 0.0909 + 0.8741 \times 0.0909 + \text{ET}(0.8741, 0.9756) \\ &\quad \times 0.0909 + \text{ET}(0.8741, 0.9756, 0.9817) \times 0.0909 \\ &= 0.037 + 0.0011 + 0.4545 + 0.0795 + 0.0773 + 0.0757 = 0.7251. \end{aligned}$$

The process of the *OWA-EUCR* is

$$\begin{aligned} \text{OEUCR}_{\text{W,EBCR}}(X) &= 0.4068 \times 0.0909 + \text{ES}\,(0.4068, 0.0473) \\ &\quad \times 0.0909 + 5.0 \times 0.0909 + 0.8741 \times 0.0909 + ES(0.8741, 0.9756) \\ &\quad \times 0.0909 + \text{ES}(0.8741, 0.9756, 0.9817) \times 0.0909 \\ &= 0.037 + 0.0405 + 0.4545 + 0.0795 + 0.0908 + 0.0909 = 0.7932. \end{aligned}$$

4   Experimentation

4.1   Experimental Settings

The maintainability is a major index to the evaluation of software trustworthiness or quality. In the *ISO/IEC 25010* software quality model, the maintainability consists of several sub-indices including "Modularity", "Reusability", "Analyzability", "Modifiability", and "Testability". We apply the evaluation of software maintainability to demonstrate the application of the mixed aggregation functions in real use. Specifically, we collect the 880 versions of Linux Kernels, which are the basis of the variety of operating systems. To combine the sub-indexes of the maintainability and the analysis of Linux kernels, we consider the 22 metrics which include four categories (They are shown in Table 3).

Table 3 Metrics description and correlation

Categories	Number	Descriptions	Correlation
Halstead complexity	O_1	Halstead Vocabulary	↓
	O_2	Halstead size	↓
	O_3	Halstead volume	↓
	O_4	Halstead difficulty	↓
	O_5	Halstead effort	↓
	O_6	Halstead errors	↓
	O_7	Halstead testing time	↓
Complexity	O_8	McCabe cyclomatic complexity	↓
	O_9	Average cyclomatic complexity	↓
	O_{10}	Maximum cyclomatic complexity	↓
	O_{11}	Essential complexity	↓
	O_{12}	Maximum nesting level	↓
Volume	O_{13}	Number of files	↓
	O_{14}	Number of functions	↓
	O_{15}	Number of lines (EXE)	↓
	O_{16}	Comments rate	↑
Volume	O_{17}	Network size	↓
	O_{18}	Number of network edges	↓
	O_{19}	Average degree	↓
	O_{20}	Clustering coefficient	↑
	O_{21}	Largest sub-graph nodes	↑

The metrics that belong to the Halstead complexity are the functions of the counts to the operators and operands in the software code. Complexity metrics and Volume metrics can be obtained by using the software tool "Understand". Mapping the methods and the dependencies of these methods into the nodes and edges of the network topology, respectively, can construct the software call network (CN) [35]. The CN is obtained by analyzing the Linux Kernels with the support of "Understand". Using the software tool "Network X" to analysis the CN, then the metrics of complex network can be acquired. The "Correlation" in Table 3 indicates the positive (↑) or negative (↓) relationship between a metric and the maintainability of software. In order to represent the metric in the next section, each metric is assigned a number ("O_i") in Table 3.

Table 4 The index system of the maintainability

Maintainability	Modularity($\omega_1 = 0.166630$)	$O_2, O_3, O_4, O_9, O_{11}, O_{12}, O_{15}, O_{17}, O_{18}, O_{20}, O_{21}$
	Reusability ($\omega_2 = 0.188758$)	$O_2, O_5, O_8, O_{12}, O_{13}, O_{14}, O_{15}, O_{18}, O_{19}, O_{20}$
	Analyzability ($\omega_3 = 0.237822$)	$O_1, O_2, O_3, O_4, O_8, O_{10}, O_{11}, O_{12}, O_{13}, O_{15}, O_{16}, O_{17}, O_{18}$
	Modifiability ($\omega_5 = 0.156509$)	$O_3, O_5, O_8, O_9, O_{12}, O_{14}, O_{15}, O_{16}, O_{18}, O_{19}, O_{20},$
	Testability ($\omega_5 = 0.250281$)	$O_1, O_2, O_3, O_4, O_6, O_7, O_8, O_{10}, O_{11}, O_{13}, O_{14}, O_{15}, O_{17}$

Based on the above arguments, we establish an index system to evaluate the maintainability of the Linux Kernels as shown in the Table 4, where every sub-index has a weight with the corresponding metrics. The weights of the sub-indexes are computed by using the method in [36]. In the experiment with n metrics in a sub-index, the weight of each metric ("O_i") is empirically set as $1/n$. For example, the sub-index "Modularity" contains 11 metrics in Table 4, and the weight of each metric is therefore 0.090909.

4.2 Implementation

Using the software tools "Understand" and "Network X", we can obtain the values of the 22 metrics in Table 3 for 880 Linux Kernels. The evaluation of maintainability for 880 Linux kernels using the proposed aggregation functions may include two steps in this experiment. Firstly, we need to pre-process the obtained data. Secondly, we evaluate the maintainability for 880 Linux Kernels using the proposed aggregation functions.

4.2.1 Data Preprocessing

First, we translate the values of each metric into the unit interval [0, 1]. If a value of the metric (O_k) is x_i, then the translating formulation is as follows:

$$u_i = \begin{cases} (x_i - L_k)/(H_k - L_k) & \text{the correlation between metric and maintainability is positive} \\ (H_k - x_i)/(H_k - L_k) & \text{the correlation between metric and maintainability is negative,} \end{cases}$$

where L_k and H_k are the low boundary and high boundary for the values of metric (O_k). In this experiment, we get the boundary of the metrics through analyzing the acquired experimental data. Such as if the values of a metric (O_k) are $\{x_i | i = 1, 2, \ldots, 880\}$ for 880 Linux Kernels, then the low boundary of metric is $L_k = mim\{x_i | i = 1, 2, \ldots, 880\}$, and the high boundary is $H_k = \max\{x_i | i = 1, 2, \ldots, 880\}$. Next, we get the concentrate regions of the metrics. The method is computed as follows:

Step 1. We partition the unit interval [0, 1] into m sub-intervals ($S_i, i = 1, 2, \ldots, m$), where m is set to 100. Next, we compute the frequencies of the values that are in sub-intervals for all metrics.

Step 2. Given two thresholds p_k and d_k of each metric O_k, we construct a set C_k that includes the intervals whose frequencies of the values are not less than p_k. The set C_k is initialized as the sub-intervals whose frequency of the values is the maximum.

Step 3. We define the distance of the sub-interval S_i to the set C_k as follows:

$$d(S_i, C_k) = \min_{S_k \in C_k} \{|S_i - S_k|\}.$$

If the distances of some intervals to the set C_k are greater than d_k, then we move these intervals from the set C_k. Finally, we combine the intervals in set C_k to get the concentrate region of metric.

4.2.2 Evaluation of Maintainability

Once the completion of the data preprocessing, the triplet $((u_i, w_k, [l_k, h_k]))$ is obtained for each value of the metric O_k, where $u_i (1 \le i \le 880)$ is the value in interval [0, 1] and, $w_k, [l_k, h_k]$ are the weight and concentrate region, respectively. Based on the index system in Table 4, the following steps are conducted to the evaluation of maintainability for 880 Linux kernels.

Step 1. We compute the aggregation results of the sub-indexes using the proposed aggregation functions, including *OWA-DEGF*, *OWA-IEGF*, *OWA-EBCR*, and *OWA-EUCR*. In order to compare the results, we also compute the averaging results of the sub-indexes.

Step 2. We compute the low and high boundaries of the sub-indexes by calculating the averaging results for low and high boundaries of the metrics to the sub-indexes.

Step 3. We compute the results of maintainability for 880 Linux Kernels by aggregating the values of the corresponding sub-indexes. Similarly, with the Step 2, we can obtain the low bounder and high boundary for the maintainability. The averaging results of the maintainability are also computed.

In addition, we apply the Einstein t-norm and t-conorm in the aggregation functions.

4.2.3 Analysis of Experimental Results

We show the main experimental results in Figs. 7, 8, and 9, where the values of the horizontal axis are the serial numbers of the Linux kernel versions with the high number indicating the high version. In Fig. 7, the "Decrease" represents the

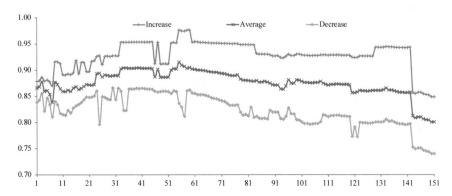

Fig. 7 Results of OWA-DEGF, OWA-IEGF, and averaging function for 151 Linux Kernels

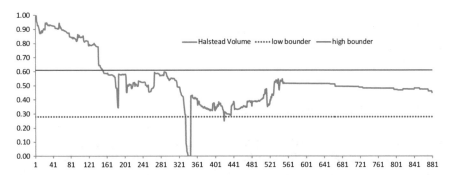

Fig. 8 Values of "Halstead Volume" for 880 Linux Kernels

aggregating results of the maintainability used in the aggregation function *OWA-DEGF*. Similarly, the "Increase" represents the aggregating results of the *OWA-IEGF*, and the "Average" represents the averaging results. In Fig. 7, there are 151 versions of the Linux Kernels with the serial number ranging from 1 to 151. The differences of three aggregating results are more obvious for these Linux Kernel versions. But for the other versions, the differences are not obvious, even though they are zero in most cases. The reason is that the most values of the metrics corresponding to these Linux Kernel versions are in *CR* of the metrics.

For example, we can obtain that most values of "Halstead Volume" belong to *CR* except for a few values starting from a Linux Kernel version, whose serial number is 152, based on the results in Fig. 8. The values of other metrics have similar characteristics. According to Fig. 7, we know that most of the aggregating results to the aggregation function *OWA-DEGF* are under the high boundary. The main purpose of the *OWA-DEGF* is to decrease the effects of the values which are outside *CR*. Most of the aggregating results to the aggregation function *OWA-IEGF* are upon the high boundary via the results in Fig. 7. On the contrary, the main purpose of the *OWA-IEGF* is to increase the effects of values which are outside *CR*. In most situations,

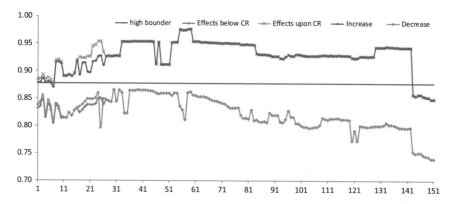

Fig. 9 Results of OWA-DEGF, OWA-EBCR, OWA-IEGF and OWA-EUCR for 151 Linux kernels

the aggregating results of the aggregation function *OWA-IEGF* are greater than the averaging results, and the results of the *OWA-DEGF* are smaller than the averaging results.

More specifically, in Fig. 9 where the "Effects below *CR*" and "Effects upon CR" represent the aggregating results of the functions *OWA-EBCR*, and *OWA-EUCR*, it is observed that the aggregating results of *OWA-DEGF* and *OWA-EBCR* are also very similar. Starting from a Linux Kernel version, the aggregating results of these two aggregation functions are same in most cases. The functions *OWA-IEGF* and *OWA-EUCR* have the same characteristics in Fig. 9. As a result, we can obtain that *OWA-DEGF* and *OWA-EBCR* can increase the effects of the values that are less than low bounder of the *CR* to the aggregating results, and the obtained results of *OWA-DEGF* and *OWA-EBCR* are very similar. Similarly, for the values that are great than upon bounder of the *CR*, the effects to the aggregating results can be increased by *OWA-IEGF* and *OWA-EUCR*, and their results are similar.

5 Conclusion

In this chapter, we propose several aggregation functions to aggregate the inputs when most values are in the concentrate region (*CR*). In order to enable the use of different aggregation operators based on the sub-regions of input domain, two binary aggregation functions are proposed, i.e., the *DEGF*, which can reduce the effects for data that lying outside of *CR*, and the *IEGF*, which can increase the effects of these data beyond *CR*. Furthermore, we extend them into multivariate functions and combine them with the popular OWA operators, which form the *OWA-DEGF*, *OWA-IEGF*, *OWA-EBCR*, and *OWA-EUCR*. The effectiveness of the proposed functions is evaluated with a case study on the maintainability of the 880 Linux Kernels. Future work will be naturally extended to implement more comprehensive systems for software fault detections as well as in combination with fuzzy systems [37, 38] designs to deal with uncertainty and impression.

References

1. Beliakov, G., Pradera, A., & Calvo, T. (2008). *Aggregation functions: A guide for practitioners.* Incorporated: Springer Publishing Company.
2. Xu, Z., & Da, Q. (2003). An overview of operators for aggregating information. *International Journal of Intelligent Systems, 18,* 953–969.
3. Zhang, H., Chen, T., Wang, Z., Chen, Y., & Chen, C. (2020). A pythagorean fuzzy preference relation approach for group decision making. In *2020 IEEE international conference on fuzzy systems (FUZZ-IEEE),* IEEE.
4. Xu, Z., & Yager, R. R. (2010). Power-geometric operators and their use in group decision making. *Fuzzy Systems, IEEE Transactions, 18,* 94–105.
5. Yager, R. R. (2012). On prioritized multiple-criteria aggregation. *Systems, Man, and Cybernetics, Part B: Cybernetics, IEEE Transactions, 42,* 1297–1305.
6. Rickard, J. T., & Aisbett, J. (2014). New classes of threshold aggregation functions based upon the tsallis q-exponential with applications to perceptual computing. *Fuzzy Systems, IEEE Transactions, 22,* 672–684.
7. Zhang, H., Zheng, Q., Wang, Z., Chen, Y., & Chen, T. (2019). Method selecting correct one among alternatives utilizing intuitionistic fuzzy preference relation without consensus reaching process. In *2019 IEEE international conference on fuzzy systems (FUZZ-IEEE).*
8. Buchanan, B. G., & Shortliffe, E. H. (1984). *Rule-based expert systems* (Vol. 3). MA: Addison-Wesley Reading.
9. Dombi, J. (1982). Basic concepts for a theory of evaluation: the aggregative operator. *European Journal of Operational Research, 10,* 282–293.
10. Yager, R. R., & Rybalov, A. (1996). Uninorm aggregation operators. *Fuzzy sets and systems, 80,* 111–120.
11. Calvo, T., De Baets, B., & Fodor, J. (2001). The functional equations of frank and alsina for uninorms and nullnorms. *Fuzzy Sets and Systems, 120,* 385–394.
12. Mesiar, R., & Komornikova, M. (1998). Triangular norm-based aggregation of evidence under fuzziness. In *Aggregation and fusion of imperfect information* (pp. 11–35), Springer.
13. Dubois, D., & Prade, H. (1984). Criteria aggregation and ranking of alternatives in the framework of fuzzy set theory. *TIMS Studies in the Management Sciences, 20,* 209–240.
14. Alsina, C., Schweizer, B., & Frank, M. J. (2006). *Associative functions: triangular norms and copulas,* World Scientific.
15. Klement, E. P., Mesiar, R., & Pap, E. (2000). *Triangular norms,* Springer.
16. Chen, T., Su, P., Shang, C., & Shen, Q. (2017). Reliability-guided fuzzy classifier ensemble. In *2017 IEEE international conference on fuzzy systems (FUZZ-IEEE)* (pp. 1–6), IEEE.
17. Su, P., Chen, T., Xie, J., Ma, B., Qi, H., Liu, J., Zhao, Y. (2020). A density and reliability guided aggregation for the assessment of vessels and nerve fibres tortuosity. *IEEE Access.*
18. Yager, R. R. (1988). On ordered weighted averaging aggregation operators in multicriteria decision making. *Systems, Man and Cybernetics, IEEE Transactions, 18,* 183–190.
19. Chiclana, F., Herrera, F., & Herrera-Viedma, E. (2001). Integrating multiplicative preference relations in a multipurpose decision-making model based on fuzzy preference relations. *Fuzzy sets and systems, 122,* 277–291.
20. Xu, Z., & Da, Q. (2002). The ordered weighted geometric averaging operators. *International Journal of Intelligent Systems, 17,* 709–716.

21. Yager, R. R., & Filev, D. P. (1999). Induced ordered weighted averaging operators. *Systems, Man, and Cybernetics, Part B: Cybernetics, IEEE Transactions, 29*, 141–150.
22. Xu, Z., & Da, Q. (2002). The uncertain OWA operator. *International Journal of Intelligent Systems, 17*, 569–575.
23. Xu, Z. (2004). *Uncertain multiple attribute decision making: methods and applications.* Beijing: Tsinghua University Press.
24. Dong, W., & Wong, F. (1987). Fuzzy weighted averages and implementation of the extension principle. *Fuzzy sets and systems, 21*, 183–199.
25. Yager, R. R. (2004). Generalized OWA aggregation operators. *Fuzzy Optimization and Decision Making, 3*, 93–107.
26. Yager, R. R. (2001). The power average operator. *Systems, Man and Cybernetics, Part A: Systems and Humans, IEEE Transactions, 31*, 724–731.
27. Yager, R. R., & Alajlan, N. (2014). Probabilistically Weighted OWA Aggregation. *IEEE Transactions on Fuzzy Systems, 22*, 46–56.
28. Menger, K. (1942). Statistical metrics. *Proceedings of the National Academy of Sciences of the United States of America, 28*, 535.
29. Schweizer, B., Sklar, A. (2011). *Probabilistic metric spaces,* Courier Dover Publications.
30. Xia, M., Xu, Z., & Zhu, B. (2012). Some issues on intuitionistic fuzzy aggregation operators based on Archimedean t-conorm and t-norm. *Knowledge-Based Systems, 31*, 78–88.
31. Beliakov, G., & Troiano, L. (2007). Fitting ST-OWA operators to empirical data. In *New dimensions in fuzzy logic and related technologies: proceedings of the 5th EUSFLAT conference, Ostrava, Czech Republic, 11-14 Sept 2007* (pp. 67–73), University of Ostrava.
32. Yager, R. R., & Troiano, L. (2005). On some properties of mixing OWA operators with t-norms and t-conorms. In: *EUSFLAT conference*, pp. 1206–1212.
33. Yager, R. R. (2005). Extending multicriteria decision making by mixing t-conorms and OWA operators. *International Journal of Intelligent Systems, 20*, 453–474.
34. Jenei, S. (1998). On Archimedean triangular norms. *Fuzzy Sets and Systems, 99*, 179–186.
35. Qu, Y., Guan, X., Zheng, Q., Liu, T., Zhou, J., & Li, J. (2013). Calling network: a new method for modeling software runtime behaviors. In *Proceedings of the second international workshop on software mining, Palo Alto, CA, USA*.
36. Zhang, H., Zheng, Q., Liu, T., Yang, Z., & Liu, J. (2013). A discrete region-based approach to improve the consistency of pair-wise comparison matrix. In *Fuzzy systems (FUZZ), 2013 IEEE international conference on IEEE* (pp. 1–7).
37. Chen, T., Shang, C., Su, P., Keravnou-Papailiou, E., Zhao, Y., & Antoniou, G., et al. (2020). A decision tree-initialised neuro-fuzzy approach for clinical decision support. *Artificial Intelligence in Medicine*.
38. Chen, T., Shang, C., Yang, J., Li, F., Shen, Q. (2019). A new approach for transformation-based fuzzy rule interpolation. *Fuzzy Systems, IEEE Transactions*. https://doi.org/10.1109/TFUZZ.2019.2949767.

Applying Fuzzy Pattern Trees for the Assessment of Corneal Nerve Tortuosity

Pan Su, Xuanhao Zhang, Hao Qiu, Jianyang Xie, Yitian Zhao, Jiang Liu, and Tianhua Chen

Abstract The tortuosity of corneal nerve fibers is correlated with a number of diseases such as diabetic neuropathy. The assessment of corneal nerve tortuosity level in *in vivo* confocal microscopy (IVCM) images can inform the detection of early diseases and further complications. With the aim to assess the corneal nerve tortuosity accurately as well as to extract knowledge meaningful to ophthalmologists, this chapter proposes a fuzzy pattern tree-based approach for the automated grading of corneal nerves' tortuosity based on IVCM images. The proposed method starts with the deep learning-based image segmentation of corneal nerves and then extracts several morphological tortuosity measurements as features for further processing. Finally, the fuzzy pattern trees are constructed based on the extracted features for the tortuosity grading. Experimental results on a public corneal nerve data set demonstrate the effectiveness of fuzzy pattern tree in IVCM image tortuosity assessment.

Keywords Fuzzy pattern tree · Computer-aided diagnosis · Tortuosity assessment · Corneal nerve.

P. Su · H. Qiu · J. Xie · Y. Zhao (✉)
Cixi Institute of Biomedical Engineering, Ningbo Institute of Materials Technology and Engineering, Chinese Academy of Sciences, Ningbo, China
e-mail: yitian.zhao@nimte.ac.cn

P. Su
School of Control and Computer Engineering, North China Electric Power University, Baoding, China

X. Zhang
School of Computer Science University of Nottingham, Ningbo, China

J. Liu (✉)
Department of Computer Science and Engineering, Southern University of Science and Technology, Shenzhen, China
e-mail: liuj@sustech.edu.cn

T. Chen
School of Computing and Engineering, University of Huddersfield, Huddersfield, UK

1 Introduction

Computer-Aided Diagnosis (CAD) systems may analyze medical images by means of machine learning algorithms in order to classify patients according to certain features and patterns that characterize a given disorder. Tortuosity is one of the important parameters of corneal neuromorphology. Changes in corneal nerve tortuosity are related to a variety of diseases such as diabetic neuropathy, retinopathy of prematurity [1], and keratitis [2]. Particularly, the decrease in subbasal corneal nerve fiber density and the increase in nerve tortuosity in diabetic patients are related to the stage or severity of peripheral neuropathy [3]. At present, IVCM images are usually graded subjectively in the band of 3–5 levels [4] with respect to the tortuosity of corneal fibers by ophthalmologists based on their clinical experiences. The grading results are highly biased, time-consuming, and difficult to be reproduced [5].

The collection of large-scale labeled images facilitates the generation of automated end-to-end CAD systems based on deep learning models, which takes raw images as inputs and then directly outputs assessment results. However, for the corneal nerve tortuosity assessment, the available data is very limited and the interpretability of the learned model is usually required in clinical settings. Conventionally, the automated tortuosity assessment may be conducted using image segmentation, feature extraction, and classification or regression [6]. For the tortuosity feature extraction, various measurements of curvilinear structures such as the angle [7], length [8], and curvature [9] are proposed in the literature. For example, Scarpa et al., contributed several algorithms for corneal nerve tortuosity evaluation and provided publicly available corneal image data sets [10]. Kim and Markoulli [11] compared different automated approaches for nerve analysis and produced tracing software to automatically segment contours in the corneal nerve and other medical applications.

Many of the existing methods focus on defining and calculating the tortuosity of individual curvilinear structures, i.e., the fiber-level tortuosity. However, as shown in Fig. 1, each IVCM image usually displays *multiple* corneal nerve fibers. The extraction of image-level tortuosity by aggregation of individual fiber-level tortuosity has substantial influence on the quality of grading. In many existing automated methods, this is implemented by the simple averaging of fiber-level tortuosity degrees or the weighted average of them with respect to fiber lengths [12–14]. However, Lagali et al. [15] conducted experiments on expert graders which showed that 'Grading only the most tortuous nerve in a given image' results in the best inter-grader repeatability [15].

Although there lacks a commonly accepted mechanism to adjudge interpretability in the tortuosity assessment task, semantics-based interpretability is typically considered when designing a method based on fuzzy set theory [16], which forms a sharp contrast with black-box systems such as deep neural networks that may achieve high performance, but with solutions difficult to comprehend. An important advantage of the proposed method lies in its readability as it supports the linguistic formulation of tortuosity measures using linguistic terms, which enable clinicians to gain insights into the assessment and trackback the result against their domain expertise.

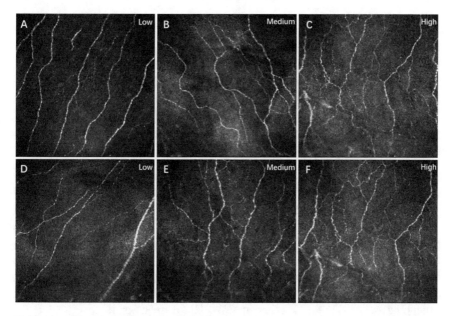

Fig. 1 Examples of IVCM Images. **A–F** are examples of corneal nerve images with different tortuosity levels of the corneal nerve tortuosity data set (columns from left to right: Grades 1–3)

This chapter proposes a Fuzzy Pattern Tree (FPT)-based pipeline for the automated grading of corneal nerves' tortuosity, whereby FPTs are generated based on the extraction of interpretable image-level features to form a robust tortuosity grading of IVCM images. Experimental analysis on a public data set demonstrates the effectiveness of fuzzy pattern tree in IVCM image tortuosity assessment.

2 Preliminaries

2.1 Triangular Norms

Triangular norm (T-norm) is a bivariate operator commonly used in the framework of fuzzy set and fuzzy logic. It can be deemed as a generalization of intersection and conjunction in fuzzy set and logic, respectively. A T-norm [17] is a mapping $T : [0, 1] \times [0, 1] \rightarrow [0, 1]$ which for all $x, y, z, x', y' \in [0, 1]$ satisfies

1. commutativity: $T(x, y) = T(y, x)$;
2. monotonicity: $T(x, y) \leq T(x', y')$, if $x \leq x'$ and $y \leq y'$;
3. associativity: $T(x, T(y, z)) = T(T(x, y), z)$; and
4. boundary condition: $T(x, 1) = x$.

A number of T-norms have been proposed in the literature such as

- the minimum T-norm: $T_{\min}(x, y) = \min(x, y)$;
- the algebraic product T-norm: $T_p(x, y) = xy$;
- the Łukasiewicz's T-norm: $T_Ł(x, y) = \max(x + y - 1, 0)$; and
- the Einstein T-norm: $T_{\mathrm{Ein}}(x, y) = xy/(2 - x + y + xy)$.

A T-conorm $S : [0, 1] \times [0, 1] \rightarrow [0, 1]$ can be defined as $S(x, y) = 1 - T(x, y)$. The first three properties of T-conorms are the same as those of T-norms while the boundary condition reads as $S(x, 0) = x$.

2.2 Ordered Weighted Averaging

Ordered Weighted Averaging (OWA) is a family of aggregation operators which may be seen as a special type of weighted averaging operator based on the ordering of inputs. The fundamental property of OWA is that the inputs are rearranged in descending order before they are integrated into a single aggregated value. Formally, a mapping $A_{owa} : \mathbb{R}^n \rightarrow \mathbb{R}$ is called an OWA [18] aggregation if

$$A_{owa}(x_1, \ldots, x_n) = \sum_{i=1}^{n} w_i x_{(i)},$$

where $x_{(i)}$ is a permutation of $\{x_1, x_2, \ldots, x_n\}$, $x_i \in \mathbb{R}$, $i = 1, 2, \ldots, n$, which satisfies that $x_{(i)}$ is the i-th largest amongst $\{x_1, x_2, \ldots, x_n\}$, and the weight $w_i \in [0, 1]$ satisfies $\sum_{i=1}^{n} w_i = 1$. The commonly used aggregation functions can be considered as special cases of OWA. For examples, the arithmetic mean can be obtained by setting $w_i = 1/n$ for all $i = 1, 2 \cdots, n$, the max by $w_1 = 1$ and $w_i = 0$ for $i \neq 1$, and the min by $w_n = 1$ and $w_i = 0$ for $i \neq n$. The aggregation behavior of OWA operators is bounded by the max and min operator as: $\min(x_1, \ldots, x_n) \leq A_{owa}(x_1, \ldots, x_n) \leq \max(x_i, \ldots, x_n)$.

In order to facilitate expert perceptions to control the aggregation behavior in OWA, the stress function has been introduced for deriving weights with measurable andness/orness and attaining interpretability [19]. Particularly, andness suggests that the aggregated result is influenced by smaller input values, while orness indicates that the aggregated result is influenced by larger input values. Let a function $h : [0, 1] \rightarrow \mathbb{R}^+$ be non-negative on the unit interval. The OWA weighting vector $W = (w_1, \ldots, w_i, \ldots, w_n)$ can then be defined as

$$w_i = \frac{h(\frac{i}{n})}{\sum_{i=1}^{n} h(\frac{i}{n})}, \tag{1}$$

such a function h is termed a stress function for OWA [19].

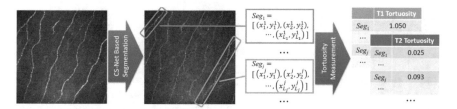

Fig. 2 The flowchart of nerve fiber segmentation and tortuosity estimation

3 Method

3.1 Nerve Fiber Segmentation and Tortuosity Measurements

The proposed pipeline of automated corneal nerve tortuosity grading consists of nerve fiber segmentation, fiber-level tortuosity estimation, image-level feature extraction, and tortuosity classification. A flowchart of nerve fiber segmentation and fiber-level tortuosity estimation is illustrated in Fig. 2.

Given an IVCM image $Img \in \mathbb{X}$, the nerve fibers $Seg_1, Seg_2, \ldots, Seg_n$ on Img are first located by a recently proposed deep learning-based algorithm termed CS-NET [20] for image segmentation. Instead of the U-Net-based convolutional neural network, the CS-NET includes a self-attention mechanism in the encoder and decoder. Both spatial and channel attention modules are utilized to further integrate local features with their global dependencies adaptively. Since the IVCM images usually contain artifacts such as small dendritic cells, which form short segments in the result of corneal nerves segmentation [5]. A simple post-process is employed to delete the segments which are shorter than 60 pixels following the running of CS-NET.

Since there is no universal agreement as to which measure to apply for when quantifying the tortuosity of nerve fibers [6], we empirically use a comprehensive range of measures, which are defined following different geometric standards. Specifically, eight fiber-level tortuosity measures (indicated as T1–T8) are calculated on each nerve fiber $Seg_i, i = 1, 2, \ldots, n$, to characterize its degrees of tortuosity by real numbers where high values represent high tortuosity. The employed fiber-level tortuosity measures are summarized in Table 1.

3.2 Image-Level Feature Extraction

It turns out that the simple averaged fiber-level tortuosity cannot provide an accurate estimation of image-level tortuosity, particularly images that consist of only a handful of highly twisted nerves among many other flat ones, but are empirically labeled highly tortuous by expert graders [10, 15]. This is attributed that high tortuosity

Table 1 Measures for Fiber-level Tortuosity Evaluation

Index	Name	Ref.	Description
T1	Arc-Chord length ratio	[21]	The ratio between curve length and the chord length
T2	Sum of curvature	[9]	The sum of absolute curvature of all points on the fiber nerve centreline
T3	Sum of squared curvature	[9]	The sum of squared curvature of all points on the fiber nerve centreline
T4	Maximum of squared curvature	[13]	The maximum of absolute curvature of all points on the fiber nerve centreline
T5	Derivative of curvature	[22]	For all points on the fiber nerve centreline, the squared derivative of curvature is summed
T6	Inflection count metric	[23]	The number of changes in sign of the curvature times arc-chord length ratio
T7	Tortuosity density	[10]	Averaged arc-chord length ratio of all portions between two consecutive inflection points
T8	Slope chain code	[7]	Averaged slope angles between two connected line segments of constant length along the curve

values from a small amount of nerves are averaged out in comparison with low tortuosity values from the majority, which leads to low estimation of tortuosity at the image level. In order to solve this problem, the OWA-based image-level feature extraction method is employed in this chapter, which aims to adjust the weights of different nerve fibers through the ordering of nerve fibers' tortuosity.

More specifically, the Ti tortuosity measurements of all nerve segments in an IVCM image will form inputs to three OWA operators whose weights are derived from the stress functions shown in Fig. 3. The piecewise stress function is defined as

$$h(z) = \begin{cases} c, & \text{if } z \in [a, b] \\ 0, & \text{otherwise} \end{cases},$$

where c is a positive constant. Stress functions of different shapes can be used to impose constraints over the distribution of weights in OWA. The values from a stress function $h(z)$ on the left side of $[0, 1]$ reflect weights associated with the larger inputs, i.e., nerve fibers with higher tortuosity degrees, whereas the values associated with the right side of the unit interval reflect the weights associated with smaller inputs, i.e., nerve fibers with lower tortuosity degrees. By setting $[a, b] = [0.1, 0.6]$, $[a, b] = [0.2, 0.4]$, and $[a, b] = [0.3, 0.5]$, three weighting vectors are generated for each IVCM image. For example, the stress function is shown in Fig. 3a indicates that

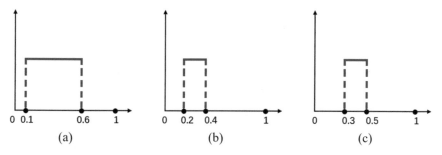

Fig. 3 Piecewise stress functions for image-level feature extraction

the nerve segments whose Ti tortuosity is higher than those of 60% segments and lower than 10% segments in the IVCM image are counted to form an image-level feature.

Formally, for each of the employed fiber-level tortuosity measure Ti and an pre-defined OWA operator $A_{owa} : \mathbb{R}^n \to \mathbb{R}$, the image-level feature Fi for an IVCM image Img is extracted by defining the value of Fi as

$$Fi(Img) = A_{owa}(Ti(Seg_1), Ti(Seg_2), \ldots, Ti(Seg_n)),$$

where $Ti(Seg_1), Ti(Seg_2), \ldots, Ti(Seg_n)$ are the tortuosity measurements of all segmented n nerve fibers in Img, estimated by Ti (See the right side in Fig. 2). Based on these image-level features extracted through OWA aggregation, a classifier can be trained to fulfill the grading of IVCM images with respect to the corneal nerve tortuosity.

3.3 Top-Down Generation of Fuzzy Pattern Trees

A FPT [24] is a binary tree with each leaf L associated with a fuzzy proposition such as 'F1 is $high$'. In this application, 'F1' is an image-level feature and '$high$' is a fuzzy linguistic term. As F1–F8 are all real-valued features, the fuzzification process is employed to transfer the real feature value $Fi(Img) \in \mathbb{R}$ to the memberships of a series of fuzzy sets such as $\{high, medium, low\}$, where μ_{high}^{Fi}, μ_{medium}^{Fi}, and μ_{low}^{Fi} are essentially membership functions defined on the domain of Fi, i.e., $\mu_{Term}^{Fi} : \mathbb{R} \to [0, 1]$, $Term \in \mathbb{T}$ and $\sum_{Term \in \mathbb{T}} \mu_{Term}^{Fi}(Fi(Img)) = 1$. For simplicity, we use $\mu_{Term}^{Fi}(Img)$ to respresent $\mu_{Term}^{Fi}(Fi(Img))$ in the following text. The fuzzy proposition is one of the basic information granules as input to grow a FPT. The root R of a FPT defines a fuzzy set, where $\mu_R(x)$ indicates the membership of instance x belongs to the fuzzy set R. A primitive FPT consists of only one node, which is both the leaf and root of the tree (as shown in Fig. 4a).

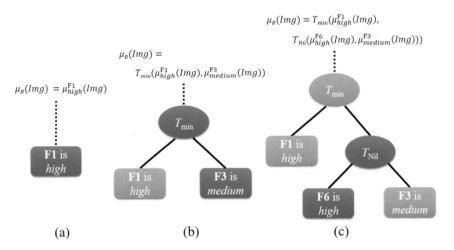

Fig. 4 An example of top-down construction of a FPT. The tree on the left is a primitive tree. The tree in the middle consists of two leaves and the root. The right one consists of three leaves, one inner node, and the root

For the classification tasks, each FPT is associated with a class label $C \in \mathbb{C}$. The fuzzy membership of instance x to the class C can be represented as $\mu_C(x)$. In crisp cases, $\mu_C(x)$ is binary such that $\mu_C(x) = 1$ indicates 'x is C', otherwise $\mu_C(x) = 0$. Given a set of IVCM images $\mathbb{X}$, for the targeting class C and an output fuzzy set at the root of a tree associated to the class R_C, the lose can be evaluated by the Rooted Mean Squared Error (RMSE) [24, 25]:

$$\mathcal{L}(C, R_C) = \left(\frac{1}{|\mathbb{X}|} \sum_{Img \in \mathbb{X}} \left(\mu_C(Img) - \mu_{R_C}(Img) \right)^2 \right)^{\frac{1}{2}}. \tag{2}$$

For a classification task involved with a set of class labels $\mathbb{C}$ ($|\mathbb{C}| > 2$), $|\mathbb{C}|$ FPTs will be constructed following the one-versus-rest decomposition such that each single tree is associated with a unique class. Given an IVCM image Img, a predication of class is made by the highest membership of Img to the $|\mathbb{C}|$ trees, i.e., arg $\max_{C \in \mathbb{C}} \mu_{R_C}(Img)$. Whereas for binary classification problems, the predictions can be made with only one single FPT via thresholding method [26].

Each inner node of a FPT is associated with a bivariate operator $O : [0, 1] \times [0, 1] \to [0, 1]$. The operator O can be T-norms, T-conorms, weighted average, and OWA with two inputs. A new fuzzy set I is formed at each inner node by applying O to its two children I_a and I_b as: $\mu_I(Img) = O(\mu_{I_a}(Img), \mu_{I_b}(Img))$. Following the same manner, a complete FPT predicts the degree of an instance belonging to its underlying class by calculating off the aggregations represented by inner nodes on the paths from the leaves to the root as shown in Fig. 4b and c.

The top-down induction of FPT [24] is an iterative process. It starts by initializing a set of candidate primitive trees, where each of all possible fuzzy propositions forms a primitive tree and is associated to each class $C \in \mathbb{C}$. The candidate tree in the next iteration $t + 1$ is an expansion of the $best T = \arg\min_{ct \in \mathbb{T}'} \mathcal{L}(C, R_C^{ct})$, where R^{ct} is the root of a candidate tree ct, by replacing one of its leaf nodes with a subtree. The subtree is made up of three nodes including the original leaf node, a new leaf node selected among the set of unused fuzzy propositions, and an inner node selected among the set of pre-defined aggregation operators. The construction process of a FPT can be briefly shown in Fig. 4 from (a) to (c), where the blue nodes are newly generated at the current iteration. These steps are repeated until a pre-defined number of iterations is reached or the improvement of $\mathcal{L}(C, R_C^{best T})$ in two successive iterations is smaller than a threshold. The details of top-down FPT construction can be found in [24].

4 Experimental Analysis

The proposed method is tested on the Corneal Nerve Tortuosity (CNT) Data Set, which is publicly available at http://bioimlab.dei.unipd.it/Data%20Sets.htm. In the CNT data set, 30 IVCM images are taken from 30 different normal or pathological subjects (diabetes, pseudoexfoliation syndrome, keratoconus). The images are manually graded into $\{high, medium, low\}$ three grades with respect to the corneal nerve tortuosity and each grade contains 10 images [27]. Eight tortuosity measures T1–T8 listed in Table 1 are employed for generating the features of tortuosity estimation. For each tortuosity measure, we use three OWA operators whose stress functions are shown in Fig. 3 to extract image-level features. In addition, the length-weighted average is also used to generate image-level feature. The Spearman's coefficients between the extracted features and the ground truth of the data set are reported in Fig. 5, where the image-level features extracted by simple average of all nerve fibers' tortuosity are provided as baselines.

The Y-axis of Fig. 5 indicates the value of Spearman's rank correlation coefficient r_s, which falls in the range of $[-1, +1]$ with the $+/-$ sign indicating the positive/negative correlation between the extracted feature and the ground truth. It can be seen from Fig. 5 that all the extracted features are positively correlated with the manual grading results. However, the results of features extracted by simple average operator are not as good as those extracted by length-weighted average and OWA operators. The average performance rank of Average, Length Weighted, and the three OWA operators over F1-F8 are 4.875, 2.875, 1.875, 2.625, and 2.750, respectively. It can be concluded from Fig. 5 that the features extracted by the OWA operators with certain stress functions are more correlated with the ground truth than those extracted by average and length-weighted average operators.

For each tortuosity measure Ti, $i = 1, 2, \cdots, 8$, we select the image-level feature which is most correlated to the ground truth among all the five extracted by different operators to form the feature space of the following FPT classification. A total number

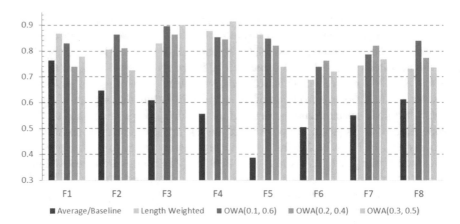

Fig. 5 Spearman's rank correlation coefficient of extracted features

of eight image-level features are selected. For example, for the tortuosity measure T2, the F2 feature extracted by the operator OWA(0.1, 0.6) is selected.

When the top-down FPT is employed to grade the tortuosity based on the extracted features, parameters that will adapt over time are the candidate set of operators connecting each inner node. In this chapter, we use the same candidate set of operators as those from [26], where T_{min}, T_p, T_L, T_{Ein} norm and the corresponding conorms, weighted average, OWA are included. In addition, there are hyperparameters such as the maximum iteration number, the pool size of bestT, and the number of fuzzy sets (i.e., terms defined to describe the value of Fi), which is set to three per feature for easy interpretation.

To demonstrate the potential variations caused by different settings of maximum iteration number and pool size, Fig. 6 is leveraged to show results by setting number of iterations from 3 to 8 and setting pool size to 15, 30, and 60. Each dot in Fig. 6 is calculated from 10 times of threefold cross-validation. Empirical results suggest that the classification accuracy of FPT is increased with the maximum number of iterations on the tested data set. However, the performances with only small variations with the paired t-test further supporting that the differences between these results are not statistically significant.

Given that r_C represents the percentage of IVCM images whose grade is C in a data set $\mathbb{X}$, TP_C, TN_C, FP_C, and FN_C are the true positives, true negatives, false positives, and false negatives for the grade C, respectively, the performance of corneal nerve tortuosity grading is evaluated based on the classification accuracy (Acc $= \frac{\sum_{C \in \mathbb{C}} TP_C}{|\mathbb{X}|}$). Following the standard performance assessment protocol employed in [6], the weighted accuracy (wAcc $= \sum_{C \in \mathbb{C}} r_C \frac{TP_C + TN_C}{|\mathbb{X}|}$), the sensitivity (wSe $= \sum_{C \in \mathbb{C}} r_C \frac{TP_C}{TP_C + FN_C}$), the specificity (wSp $= \sum_{C \in \mathbb{C}} r_C \frac{TN_C}{TN_C + FP_C}$), positive predicted value (wPpv $= \sum_{C \in \mathbb{C}} r_C \frac{TP_C}{TP_C + FP_C}$), and negative predictive value (wNpv $= \sum_{C \in \mathbb{C}} r_C \frac{TN_C}{TN_C + FN_C}$) are also employed to evaluate the performance based

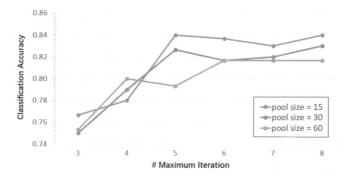

Fig. 6 Accuracy of FPT classification with different parameters

Table 2 Corneal Nerve Tortuosity Grading Performance Evaluation

	Acc	wAcc	wSe	wSp	wPpv	wNpv
C4.5 decision tree	0.7867	0.8400	0.7867	0.8933	0.7902	0.8970
3 nearest neighbors	0.8467	0.8850	0.8467	0.9233	0.8562	0.9265
Support vector machine	0.8133	0.8600	0.8133	0.9067	0.8156	0.9116
Top-down FPT	0.8400	0.8800	0.8400	0.9200	0.8455	0.9205

on four classification methods and their results are reported in Table 2. Three well-known classification models: the C4.5 decision tree, nearest neighbor classification, and support vector machines with linear kernel are employed as baselines. All these models are implemented in the WEKA machine learning framework [28] with the parameters set identical to [24]. For the FPT, the maximum iteration number is set to 5 and the pool size is set to 15. The averaged results based on 10 times of threefold cross-validation are reported in Table 2.

In general, most of the IVCM images are located far from the classification boundaries in the feature space, while only five or six instances are difficult to be correctly classified for all the tested classifiers. As reflected in Table 2, by applying the FPT, the grading results are better than those obtained by C4.5 decision tree and support vector machine using linear kernel. Despite such differences are not statistically significant as evaluated by the paired t-test, such performance competent to popular machine learning benchmarks still demonstrates the effectiveness of fuzzy pattern tree in IVCM image tortuosity assessment.

It is worth noticing that not only the accuracy of tortuosity grading, but also the interpretability of features and transparency of inferences are considered in practice. Compared to the decision tree, which could generate numerous rules, the fuzzy pattern tree is more compact that directly reveals the expression (pattern) of which and how the features are integrated to approach one certain tortuosity grade. However, the top-down induction of fuzzy pattern tree is more time-consuming compared with the other three-tested classifiers, especially if the pool size is set to a large number. This is mainly due to that in each iteration, the algorithm needs to check a large number of alternative modifications, which slows down the search process. Despite the time-consuming searching process in the training of fuzzy pattern tree, the online testing speed of fuzzy pattern tree is similar to that of conventional decision trees and meets the time efficiency requirement of CAD systems.

5   Conclusion and Future Work

This chapter proposes a pipeline for the grading of corneal nerve tortuosity based on FPT. The proposed work is verified on a public real-world data set in comparison to other classification models, demonstrating the effectiveness of the proposed method. Whist promising, the proposed pipeline could be extended to cope with a broader range of CAD tasks where explicit feature extraction and interpretable classification are required.

Acknowledgements This work was supported by a Project funded by China Postdoctoral Science Foundation (2019M652156), National Natural Science Foundation of China (61906181), and Zhejiang Provincial Natural Science Foundation China (LQ20F030002).

References

1. Heneghan, C., Flynn, J., O'Keefe, M., & Cahill, M. (2002). Characterization of changes in blood vessel width and tortuosity in retinopathy of prematurity using image analysis. *Medical Image Analysis, 6*(4), 407–29.
2. Kurbanyan, K., Hoesl, L., Schrems, W., & Hamrah, P. (2012). Corneal nerve alterations in acute Acanthamoeba and fungal keratitis: An in vivo confocal microscopy study. *Eye, 26*, 126.
3. Messmer, E. M., Schmid-Tannwald, C., Zapp, D., & Kampik, A. (2010). In vivo confocal microscopy of corneal small fiber damage in diabetes mellitus. *Graefes Archive for Clinical & Experimental Ophthalmology, 248*, 1307–1312.
4. Oliveira-Soto, L., & Efron, N. (2001). Morphology of corneal nerves using confocal microscopy. *Cornea, 20*(4), 374–384.
5. Mehrgardt, P., Zandavi, S. M., Poon, S. K., Kim, J., Markoulli, M., & Khushi, M. (2020). U-net segmented adjacent angle detection (USAAD) for automatic analysis of corneal nerve structures. *Data, 5*, 37.
6. Annunziata, R., Kheirkhah, A., Aggarwal, S., Hamrah, P., & Trucco, E. (2016). A fully automated tortuosity quantification system with application to corneal nerve fibres in confocal microscopy images. *Medical Image Analysis, 32*, 216–232.
7. Bribiesca, E. (2013). A measure of tortuosity based on chain coding. *Pattern Recognition, 46*, 716–724.
8. Grisan, E., Foracchia, M., & Ruggeri, A. (2008). A novel method for the automatic grading of retinal vessel tortuosity. *IEEE Transactions on Medical Imaging, 27*, 310–319.
9. Hart, W. E., Goldbaum, M. H., Kube, P., & Nelson, M. (1999). Measurement and classification of retinal vascular tortuosity. *International Journal of Medical Informatics, 53*(2–3), 239–52.

10. Scarpa, F., Zheng, X., Ohashi, Y., & Ruggeri, A. (2011). Automatic evaluation of corneal nerve tortuosity in images from in vivo confocal microscopy. *Investigative Ophthalmology Visual Science, 52*(9), 6404–6408.
11. Kim, J., & Markoulli, M. (2018). Automatic analysis of corneal nerves imaged using in vivo confocal microscopy. *Clinical and Experimental Optometry, 101*, 147–161.
12. Annunziata, R., Kheirkhah, A., Aggarwal, S., Cavalcanti, B., Hamrah, P., & Trucco, E. (2014). Tortuosity classification of corneal nerves images using a multiple-scale-multiple-window approach. In *Proceedings of the Ophthalmic Medical Image Analysis*, pp. 113–120.
13. Annunziata, R., Kheirkhah, A., Aggarwal, S., Cavalcanti, B. M., Hamrah, P., & Trucco, E. (2016). Two-dimensional plane for multi-scale quantification of corneal subbasal nerve tortuosity. *Investigative Ophthalmology & Visual Science, 57*, 1132–1139.
14. Ramos, L., Novo, J., Rouco, J., Romeo, S., Álvarez, M. D., & Ortega, M. (2018). Retinal vascular tortuosity assessment: Inter-intra expert analysis and correlation with computational measurements. *BMC Medical Research Methodology, 18*, 1–11.
15. Neil, L., Enea, P., Patel, D. V., Mcghee, C. N. J., Pedram, H., Ahmad, K., Mitra, T., Petropoulos, I. N., Malik, R. A., & Paaske, U. T. A. (2015). Focused tortuosity definitions based on expert clinical assessment of corneal subbasal nerves. *Investigative Ophthalmology & Visual Science, 56*, 5102–5109.
16. Chen, T., Shang, C., Su, P., & Shen, Q. (2018). Induction of accurate and interpretable fuzzy rules from preliminary crisp representation. *Knowledge-Based Systems, 146*, 152–166.
17. Schweizer, B., & Sklar, A. (2011). *Probabilistic metric spaces*. Courier Dover Publications.
18. Yager, R. (1988). On ordered weighted averaging aggregation operators in multicriteria decisionmaking. *IEEE Transactions on Systems, Man and Cybernetics, 18*, 183–190.
19. Yager, R. (2007). Using stress functions to obtain OWA operators. *IEEE Transactions on Fuzzy Systems, 15*, 1122–1129.
20. Mou, L., Zhao, Y., Chen, L., Cheng, J., Gu, Z., Hao, H., Qi, H., Zheng, Y., Frangi, A., & Liu, J. (2019). CS-Net: Channel and spatial attention network for curvilinear structure segmentation. In *Medical Image Computing and Computer Assisted Intervention—MICCAI 2019, Proceedings*, pp. 721–730.
21. Lotmar, W., Freiburghaus, A., & Bracher, D. (1979). Measurement of vessel tortuosity on fundus photographs. *Albrecht Von Graefes Archiv Für Klinische Und Experimentelle Ophthalmologie, 211*, 49–57.
22. Patasius, D. J. M., Marozas, V., & Lukosevicius, A. (2005). Evaluation of tortuosity of eye blood vessels using the integral of square of derivative of curvature. In *Proceedings of the 8th European Medical and Biological Engineering Conference (EMBEC05)*.
23. Bullitt, E., Gerig, G., Pizer, S. M., Lin, W., & Aylward, S. R. (2003). Measuring tortuosity of the intracerebral vasculature from MRA images. *IEEE Transactions on Medical Imaging, 22*, 1163–1171.
24. Senge, R., & Hüllermeier, E. (2011). Top-down induction of fuzzy pattern trees. *IEEE Transactions on Fuzzy Systems, 19*, 241–252.
25. Huang, Z., Gedeon, T. D., & Nikravesh, M. (2008). Pattern trees induction: A new machine learning method. *IEEE Transactions on Fuzzy Systems, 16*, 958–970.
26. Senge, R., & Hüllermeier, E. (2015). Fast fuzzy pattern tree learning for classification. *IEEE Transactions on Fuzzy Systems, 23*, 2024–2033.
27. Fabio, S., Xiaodong, Z., Yuichi, O., & Alfredo, R. (2011). Automatic evaluation of corneal nerve tortuosity in images from in vivo confocal microscopy. *Investigative Ophthalmology & Visual Science, 52*, 6404.
28. Eibe Frank, M. A. H., & Witten, I. H. (2016). *The WEKA Workbench. Online Appendix for Data Mining: Practical Machine Learning Tools and Techniques*, (4th ed.). Morgan Kaufmann.

A Mamdani Fuzzy Logic Inference System to Estimate Project Cost

Daniel Helder Maia and Arjab Singh Khuman

Abstract The precision and reliability of estimations of project costs are essential, especially in significant cooperation. The level of uncertainty when estimating projects can cause issues down the line during a project. For generations, humans are more often than always in a predicament where estimation for a project size or cost appears to be complicated. The methodology adopted in this research included using the literature to review the topic of project estimation and explore the use of fuzzy logic in order to define an initial fuzzy system. The development of a system to estimate project costs is based on findings from the literature. This work seeks to demonstrate the benefits of using fuzzy logic in estimating the cost for business. Analysis of the results attained during testing and research shows that the system could be beneficial for estimating the cost of projects. The results show that the system can produce an appropriate result when estimating project cost. The study concludes that there is still room for improvement and that further development and testing could lead to improvements; however, the current system gives a foundation for further development such that the system can be put to use in a real-world situation. Whether it is for business or personal circumstances where any or most cases, cost estimation is required.

Keywords Cost estimation · Time estimation · Size estimation · Effort estimation · Expenses estimation

1 Introduction

How can businesses use a fuzzy inference system (FIS) to estimate how much a particular project should cost based on variables such as, time, size, expenses and travel time? Over the years, project managers, either underestimate or overestimate their projects. As humans, expecting to accurately estimate 100% the price of a particular project is based on how long it will take to complete by date/time, travel

D. H. Maia · A. S. Khuman (✉)
De Montfort University, Leicester, UK
e-mail: arjab.khuman@dmu.ac.uk

© Springer Nature Switzerland AG 2021
J. Carter et al. (eds.), *Fuzzy Logic*,
https://doi.org/10.1007/978-3-030-66474-9_10

time, etc. This approach is fundamentally impossible. There have been many research papers, and researchers who have tried to calculate ways for specific tasks, but yet it has become almost virtually impossible to estimate a particular outcome accurately.

Due to unexpected changes of circumstances to a project, it can be difficult to estimate what should be completed when by a particular date, however, we still cannot be sure this will happen however the prediction is made. In essence, as an example, if we have project A with time, size, travel time, expenses and effort, as humans we can measure and predict that the project can/should be completed within say 6 months and cost X, yet the project completion may well end up taking 12 months costing the company more than the profit from the project. This principle applies to any other type of estimation, including the expenses of a project and predicting travel time spent. This is an ongoing issue that project managers face on a day-to-day basis, in fact, everyone faces it in our personal lives and personal projects.

With this context in mind, how can we use a Mamdani Fuzzy Logic Inference System, which first appeared in 1965 [1] to assist project managers to better estimate projects based on variables? We will incorporate this idea and build a FIS system which will allow and assist project managers to carefully estimate a project cost to get an even closer accurate estimation of projects.

2   Literature Review and Motivation

To understand the role and implementation of fuzzy logic in estimating projects, firstly we need to look at one of the most widely used models such as the Constructive Cost Estimation Model (COCOMO) [2]. This model enables software developers to estimate, time, effort and cost based on three categories. These are organic, semi-detached and embedded. The organic category is considered a group of experienced individuals in a small team with a small level of complexity, whereas, semi-detached aimed at experienced and new individuals in small to medium-size teams with a comparatively larger complexity than organic. Finally, embedded is aimed at a large experienced team with high complexity.

There is evidence in research demonstrating that fuzzy logic gives better software project estimates than the traditional COCOMO model [2]. Likewise, Reddy and Raju [3] agree that COCOMO is less efficient in its estimation compared to using fuzzy logic. Reddy and Raju [3] and Attarzadeh and Hock Ow [2] all agree that it yields uncertainty in the output, which leads to errors. It appears that fuzzy logic-based estimation models are more appropriate when dealing with ambiguous and unreliable information.

There is a general agreement from the research conducted and collected: it appears that the FIS is more accurate compared to the COCOMO model. However, there appears to be no further research on change of circumstances during the project process. Even though compared to other models, further research in depth to determine for which are suitable for further careful estimation as the design of the FIS.

Building on the idea that a FIS can be used to calculate estimation carefully, we need to look at the benefits of fuzzy logic in business and whether businesses would benefit. As we have seen thus far, it can be beneficial to implement fuzzy logic into different aspects of businesses. Specifically, in estimating projects, however, what about other aspects of business such as purchasing items?

Bezděk [4] collected information by approaching sales assistants and other employees from a Baumax store in Zlin. After asking 30+ questions, the outcome was that they overwhelmingly found that the staff were in favour of using fuzzy logic in their business. Furthermore, using the idea implemented in the paper, he/she found in three simple steps the best lawnmower identified for the customer and claimed the use of fuzzy logic therefore accelerates customer service. In accordance with the evidence outlined above and the results from the questionnaire, it appears fuzzy logic can be used and implemented in many areas of business. However, what has not been considered is whether or not fuzzy logic can alternatively cause a complication using this approach.

This next section provides a general discussion of how businesses encounter failed projects due to sparse estimation or in uncommon cases overestimated. Looking at how or why projects fail will allow us to be in a better position to design a fuzzy logic system to prepare for project estimation.

Doloi [5] discusses the shortfalls of traditional practices of estimation in which it plays a part in the failure of projects. Although he/she explains that estimation has an impact on the failure of projects, the author also stated that political, economic, technical and behavioural perspectives play a significant role in business case development. Although useful, there was not enough research found to get a concise cause of issues related to project failures. It would be interesting to know at which point projects fail and why precisely.

We will continue to look at some of the main points which are vital in the estimation process, which includes but is not limited to size, cost and time. Research was conducted on each of the variables and how they can be optimised with the implementation of fuzzy logic.

Several reports have shown that to estimate project quality or any aspect of the project, the critical inputs are time, cost and size. Studies conducted by Kharola and Singh [6] used time and cost as inputs to determine the quality of a project, by merely placing time and cost into a FIS and using the quality as the output. Similarly, Marandi and Khan [7] conducted a similar study, with minor differences in the inputs. For instance, Marandi and Khan [7] took consideration of the cost of failure into account, along with the general time and cost. Also, Reddy and Raju [3] both concentrated on the size of the project in conjunction with the COCOMO methods.

These approaches show that it is viable to use the variables of time, cost and size to create a custom FIS for project estimation ultimately. However, not all research conducted shows any debate on the use of different membership functions. Marandi and Khan [7] conducted the tests using only the trapezoidal and triangular membership functions. Therefore, further research is necessary for investigating additional membership functions and applying them to collect further samples to analyse further.

3 The System

3.1 System Design Overview

The proposition outlined below for this system is to allow businesses the ability to have their projects' cost estimated based on input variable values passed to the system. As we all know in project management and estimating projects, in general, cost is amongst one of the most significant issues all businesses encounter. We will look into a FIS designed using the same principles already researched in the literature review in Sect. 2, along with the understanding of the problem.

The view of the problem, as mentioned, is that businesses are unable to estimate project costs. Businesses tend to underestimate or overestimate. Perception of the solution is to use Fuzzy Logic to minimise estimation from a crisp binary result of either overestimated or underestimated and allowing fuzzy logic to give a vague estimation on the project cost.

This system has been split into two systems, starting with essential parts of estimation for any project which includes the time and size of the project. The outcome of those inputs will provide us with the effort required for the project. It is followed by using the output of that system as the input for the second inference system. Since the FIS not only has an output but also is used as an input for the second FIS, we add the expenses and travel time as the additional inputs to the second FIS. We now have the two inputs being expenses, travel time and the output from the first FIS known as effort. These inputs will make up the second FIS which will then give the output of the project cost.

A full visual diagram of the system can be seen in Fig. 1, how each of the components is modelled, to produce the full concept of the system. The system is broken into two parts to get a better understanding of how each of the parts plays in their system.

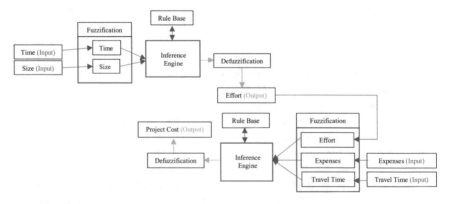

Fig. 1 Full fuzzy inference system

3.2 Project Effort Fuzzy Inference System

Overview

With the first inference system, we will consider the project effort. In this section, we will have two inputs and one output, and each of the inputs will allow us to calculate the output for the required effort for a particular project. As seen in Fig. 1, the system consists of two inputs, time and size.

Justification

Each of the variables was chosen based on research papers conducted in the literature review. How long it takes to complete the project truly depends on the size and the time it takes to complete. With this information, we will calculate the effort to pass through to the second inference system.

With regards to time, the decision to use a range of 0–100% was based on research conducted in the literature review. However, we can consider that a small percentage represents a meagre amount of time for the project to be completed, whereas the higher the percentage, the more time is available for the project. Kharola and Singh [6] used the same range values of 0–100% within their research. The interval values for this work were picked according to those used by Kharola and Singh [6].

The size variable will also have a range of 0–100%. The same principle applies as the time variable. The smaller the percentage, the smaller the size of the project and vice versa, the larger the percentage, the bigger the project. Even though the specified range was correctly picked, other researchers such as Sharma and Verma [8] used the same values in their research.

The range for the effort was chosen based on a paper by Sharma and Verma [8], where they specified the range of 0–1000 to represent the size. It would be more appropriate to have a range of 0–100% for not only simplicity but also the ease of measurement in general. The intervals, on the other hand, were chosen based on simplicity since it will not make much of a difference or effect to the system from using either 1000 or 100.

Fuzzification

The three types of membership functions used vary from the Triangular to the Trapezoidal and Gaussian membership functions.

The input for time utilises five membership functions which all five use the Triangular membership function. The decision to use the membership functions was taken based on Kharola and Singh [6].

The size input has been developed with six Gaussian membership functions. These membership functions were used based on research conducted by Sharma and Verma [8], however, since a new interval was added another membership function was also added based on the rest of the membership functions used for simplicity and consistency. The output utilises six Triangular membership functions.

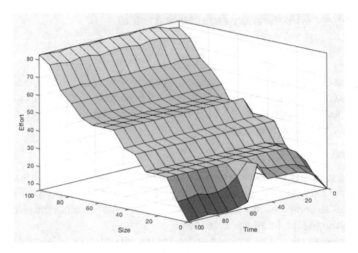

Fig. 2 Surface view of the project effort fuzzy inference system

Fuzzy Rule Base

Forty-one rules were created using the two inputs of time and size, over time during testing it will be amended to identify the rules to keep, add or remove. The fuzzy rule base infers the output based on the two input variables. The rule base has utilised the modus ponens as such

$$Antecedent - IF\ x\ is\ A\ THEN$$
$$Consequent - y\ is\ B$$

Defuzzification

The defuzzification is part of the process in which it produces a quantifiable result in crisp logic format. The first part of the FIS, defuzzification has been kept using the centroid to calculate the centre area of the method, which is commonly called the centroid. Then the x coordinate of the centroid is the defuzzied value (Fig. 2).

3.3   Project Cost Fuzzy Inference System

Overview

In the second fuzzy inference system, we will look at how the system will be completed. The output of the first inference system, Effort, will then be used with the other two inputs, expenses and travel time to complete the full FIS inputs; the final output will produce the project cost. The second inference system has been designed, which utilises the output from the first project effort inference system to continue through the system to calculate the project cost. The project cost inference system has been designed to utilise the output from the project effort inference system to continue through the system to calculate the project cost.

Justification

As previously mentioned in the first inference system, why the decision to use the measurements of 0–100% and the values of the interval will be the same used for this new inference system.

The project cost is going to be measured in the same way as the other inputs of 0–100%. The choice of these measurements is due to maintaining consistency with the other inputs and based on the research conducted by Kharola and Singh [6]. In contrast, Attarzadeh and Hock Ow [2] used the measurement of 0–1.25, however, the preference of measurement chosen here for simplicity and consistency remains at 0–100%.

In general, time spent travelling for business will always differ based not only on a project but also by location. An initial idea for this is to use a range of 48 h. Nowadays, people can travel to the other side of the world within 24 h. The other 24 h are provided for leeway for travel disruptions and flight connections. However, since some business individuals travel for a living without a home base, it would be difficult to find an accurate representation in terms of exact hours for such cases.

In which case, a range of 0–100% is again the most appropriate range to use on the basis that we can say that anything above say 70% is viewed as a very long time spent travelling, likewise on the other side of the spectrum where anything less than 20% can be seen as short travel time or 0% seen as no travel time. This way we can adequately represent all cases.

Since measuring expenses is similar to that when dealing with cost, the judgement of using the same range and intervals as the project cost appears to be the most appropriate to use.

Fuzzification

Part of this second FIS has made use of two different membership functions, and this includes the Triangular and the Trapezoidal ones. There are three inputs and one output, and we will look into which membership functions are used for those variables. The first input we will look into is the output from the first inference system, which is project effort. The result of the first inference system will feed into this second inference system. In this case, the same membership functions have been used, which is the Triangular membership function. The second input is expenses, based on research conducted earlier in the report, having used the same membership function as the output (project cost) since expenses are calculated the same way very much. The third input is travel time. A mixture of Triangular and Trapezoidal membership functions was used. The intervals short, medium and long have been used for the Triangular functions whereas concise and very lengthy were used for the trapezoidal function. Finally, the project cost, which is the output, makes use

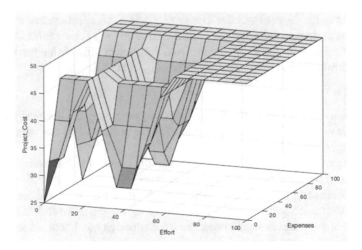

Fig. 3 Surface view of the project cost fuzzy inference system

of the triangular membership function. This decision was made again on research conducted earlier in the paper Kharola and Singh [6].

Fuzzy Rule Base

Thirty-six rules were created for the initial development of the system. The output from the first inference system is used as an input for this inference system, with the addition of two other inputs of travel time and expenses for the project cost calculation. For example, one of the rules is if the effort is very low and expenses are very low and travel time is very short, then the project cost will be very low.

Defuzzification

The Largest of Maximum (LOM) will be used initially to get the most significant outcome from the output. It can be seen later that tests were conducted with other defuzzification methods to identify which method best suited the system. After creating the surface viewer, further testing was required. See Fig. 3 of the first fuzzy inference system, the surface viewer.

4   Experimental Design and Evaluation

The tests were conducted using MATLAB version R2019b. The three most common membership functions were used; these are the Triangular, Trapezoidal and the Gaussian functions. The process of identifying the best suitable membership function for each of the inputs and outputs was tricky. However, those that have been used seem the most suitable at this time and following research conducted during the literature review.

Test One: Rules Adjustments

Changes

Firstly, we will look at going over a small set of rules to ensure the most appropriate defuzzification method to use, which will give us a steppingstone onto further tests, attempting to ensure that the rules for very low time and very large size, in particular, provide an appropriate result before proceeding with additional rules and tests.

Expected Outcome

Expecting only small differences when testing across all the defuzzification methods and believing there will be a suitable method upon completion of this test.

Results of the test

The changes did not produce accurate results across the two operators and all the defuzzification methods. There were only two output results that stood out, and it was felt that the AND operator was not the most suitable one to use for this particular system since the results were inconsistent (Table 1).

Defuzzification

The five defuzzification methods were used to get a crisp output for this inference system. Each of the defuzzification methods has produced results against all the sample data. The produced results varied, however, in some instances; centroid (rather than LOM) produced the closest prediction.

Analysis

During the tests, it was found that all the defuzzification methods produced some relevant results; however, many were not relevant. The decision was taken to alter the naming of the fuzzy sets for both size and time to make it easier to identifying and implementing the rules. Changing the membership functions for the fuzzy set size to use the Triangular membership function may help. Minor changes were made to the membership functions for time. The fuzzy sets were changed because, in the instance of size, the original naming convention did not appear to be most appropriate. In reality, when we measure size, we think small, medium, large work well just as we would when shopping for clothes which is why the decision to rename it was made. The time fuzzy set names were kept the same with one minor change and removing the Extremely High (EH) since it was no longer required and would have no effect since the rules would not make use of it.

Test Two: Re-test Test 1 Changes

Changes

Membership functions and their values have been changed for both inputs and the effort output. The fuzzy sets have also been modified, as mentioned in the test one analysis. We added further rules to this test, however, not expecting the outputs to change too much.

Table 1 Represents AND and OR operators for test one

AND operator (%)

Time	Size	Centroid	Bisector	MOM	SOM	LOM
90	5	7.16091954	7	4	0	8
5	90	92.8609338	93	96.5	93	100
1	90	92.8609338	93	96.5	93	100
90	1	7.16091954	7	4	0	8
99	1	6.357552581	6	0	0	0
1	99	93.64244742	94	100	100	100
10	90	92.83908046	93	96	92	100
90	10	7.16091954	7	4	0	8
50	25	50	50	50	50	50
25	50	50	50	50	50	50

OR operator (%)

Time	Size	Centroid	Bisector	MOM	SOM	LOM
90	5	6.351945602	6	0	0	0
5	90	93.4	94	98	96	100
1	90	93.64244742	94	100	100	100
90	1	6.351945602	6	0	0	0
99	1	6.351945602	6	0	0	0
1	99	93.64245073	94	100	100	100
10	90	92.8609338	93	96.5	93	100
90	10	6.904077562	6	3	0	6
50	25	9.5	9	9.5	0	19
25	50	9.909172282	10	9.5	0	19

Expected Outcome

There was an expectation of improved results due to the alteration to the membership functions replacing the old fuzzy set names with something more appropriate. The tests used the same sample data as before to remain consistent across all the tests.

Results of the test

There were drastic changes when using the OR operator, along with all the defuzzification methods compared to the last test. They seem far from appropriate compared to the previous results. However, the results for using the AND operator produces the exact results expected when using the default defuzzification method of Centroid (Table 2).

Table 2 Represents AND and OR operators for re-testing test one

AND operator (%)

Time	Size	Centroid	Bisector	MOM	SOM	LOM
90	5	5.3030303	5	3	0	6
5	90	93.9230769	94	95	90	100
1	90	93.9230769	94	95	90	100
90	1	5.3030303	5	3	0	6
99	1	4.70588235	4	0.5	0	1
1	99	95.2941176	96	99.5	99	100
10	90	93.9230769	94	95	90	100
90	10	6.07692308	6	5	0	10
50	25	50	50	50	50	50
25	50	50	50	50	50	50

OR operator (%)

Time	Size	Centroid	Bisector	MOM	SOM	LOM
90	5	50	50	50	0	100
5	90	50	50	50	0	100
1	90	50	50	50	0	100
90	1	50	50	50	0	100
99	1	49.9774368	50	0	0	0
1	99	50	50	50	0	100
10	90	50	50	50	0	100
90	10	46.9514563	47	3	0	6
50	25	37.5	38	37.5	25	50
25	50	62.5	63	62.5	50	75

Defuzzification

All the same five defuzzification methods were used to generate and get a crisp output.

Analysis

Across all the defuzzification methods and operators, the most appropriate from a simple test is when expecting the time to be around the 10% region, therefore very low, and the size expected to be around 90%; the expected effort would be around 90%.

The same goes when the time is very high at around 90% and the size at around 10%. The expectation is to have the effort to be lower. With Centroid, those exact results are produced making this the best defuzzification method for this system along with using the AND operator.

Test Three: Weight Value Changes

Changes

The weight will be amended to the rules to see if the weight changes will have any effect on the output. The weight values will be amended to 1, 0.75, 0.5, 0.25 and 0.

Expected outcome

Not expecting the results to change much from the most recent results produced. However, there will be minor changes to the results but not much to warrant any significant further changes to the system.

Results of the test

As expected, there was not much of a change in the results from the outcome of the tests. Much of the results were comparable to the previous tests.

Defuzzification

Only the centroid defuzzification method was the most important for this test, however, it created a table of comparisons against the other defuzzification methods. The operator has not changed and will remain using AND since the OR was weak in the previous tests.

Analysis

As expected from the results, the weight did not make much of a change to the results in which case the weight will remain at 1 for the remainder of the tests. The result was the same across all the defuzzification methods.

Test Four: Prod Implication and Sum Aggregation Methods

Changes

During this test, the implication and aggregation will be altered for the inference system, to see if this has any effect on the result. The test will include all the defuzzification methods across all the weight changes which have been tested as before.

Expected outcome

With the expectation to have significant changes to the output; more specifically, the result will have a considerable significant difference.

Results of the test

The prod implication method test was again used against the same sample data and modified rules as the previous test. The sum aggregation method, as investigated, found the results of using the prod implication method for this inference system along with a comparison to all the defuzzification methods to be beneficial (Table 3).

Table 3 Prod Implication and Sum Aggregation methods for AND and OR operators for test four

AND operator (%)						
Time	Size	Centroid (weight: 1)	0.75	0.5	0.25	0
90	5	4.666666667	4.66666667	4.6666667	4.66666667	50
5	90	95.33333333	95.3333333	95.333333	95.3333333	50
1	90	95.33333333	95.3333333	95.333333	95.3333333	50
90	1	4.666666667	4.66666667	4.6666667	4.66666667	50
99	1	4.666666667	4.66666667	4.6666667	4.66666667	50
1	99	95.33333333	95.3333333	95.333333	95.3333333	50
10	90	95.33333333	95.3333333	95.333333	95.3333333	50
90	10	4.666666667	4.66666667	4.6666667	4.66666667	50
50	25	50	50	50	50	50
25	50	50	50	50	50	50
OR operator (%)						
Time	Size	Centroid (weight: 1)	0.75	0.5	0.25	0
90	5	50	50	50	50	50
5	90	50	50	50	50	50
1	90	50	50	50	50	50
90	1	50	50	50	50	50
99	1	49.82972334	49.8297233	49.829723	49.8297233	50
1	99	50	50	50	50	50
10	90	50	50	50	50	50
90	10	45.64941129	45.6494113	45.649411	45.6494113	50
50	25	37.5	37.5	37.5	37.5	50
25	50	62.5	62.5	62.5	62.5	50

Defuzzification

All five defuzzification methods were tested as before, including centroid, bisector, MOM, LOM and SOM. This was also including the testing against the two operators AND and OR.

Analysis

In this instance, prod did not show much improvement to the system across all of the defuzzification methods, except the Centroid and Bisector where there was some minuscule improvement; with regards to the other defuzzification methods, they made the results much worse. Similarly, the use of sum made slight changes, again to the Centroid and Bisector defuzzification methods—there were small changes in results and like prod, the other defuzzification methods did not show an improvement in the results.

4.1   Project Cost Fuzzy System Tests

Test One: Rules Adjustments

Changes

As we did in the first inference system, we will first look at the current rule base and reduce the number of rules. Currently, there are 36 rule entries. The goal of this test is to reduce the number of rules and to ensure the inputs and outputs have the most appropriate membership functions associated with their fuzzy set, and these rule bases were altered. It is tested across all the defuzzification methods with both the AND and OR operators to determine which will be the most appropriate.

Expected outcome

Expected to have a much-reduced rule base list and notice that the membership function used may need altering as it did in the first inference system. This will provide a more accurate result on the test data.

Results of the test

After reducing the number of rules from 36 to 15, it made a significant difference in the system outcome, and it is still providing unexpected results. Not only is it suitable from a testing perspective since the system will be quicker to run, but also the rule base is sufficient to carry on with the tests further. After looking at the membership functions, alteration is required, especially for the output and the expenses input (Table 4).

Defuzzification

The five defuzzification methods were tested across the test data with the minimised rule base along with the two operators to get a crisp output for this inference system.

Table 4 Results from the data against the rule base

Project effort	Expenses	Travel time
5.303030303	100	100
93.92307692	25	50
93.92307692	50	25
5.303030303	10	0
4.705882353	75	25
95.29411765	20	75
93.92307692	5	5
6.076923077	30	50
50	50	50
50	35	0
50	0	0

Analysis

During this test, as expected, the membership functions need to be altered for both the project cost output and the expenses input. The travel time input appears to be just as good for the time being, however, it cannot rule out future changes at this moment in time. With regards to the results, it is currently inconclusive since the membership functions will need altering.

As seen in the result table, this gives inconclusive results, for example, when using the AND operator in particular for Centroid, the project effort returns 95% from the first inference system, expenses 5%, and travel time at 5% I would then expect the result to be around 80% or 85%, however, in this case, the result was 90%.

Test Two: Prod Implication and Sum Aggregation Methods

Changes

The changes made will cater to both the implication method and aggregation. When using the implication method, it will use the prod and the aggregation will use the sum. These two methods will test across all defuzzification methods and weights to decide on the most appropriate one to use.

Expected outcome

Expecting to get similar results as before for the first inference system, whereas most of the minor changes will be for the OR operator across all the defuzzification methods.

Results of the test

As expected, the results appeared to make the most difference using the OR operator. However, the implication method across all the weights based on the chosen defuzzification method caused some changes. Using the prod implication method based on the chosen defuzzification method, the values stayed consistent, meaning that no matter what the weight is the results will remain consistent. On the other hand, while using the sum aggregation method, their minimal differences.

Defuzzification

All five defuzzification methods will be tested against as before including centroid, bisector, MOM, LOM and SOM so that it will allow us to collect an exciting result set to compare against to make a better decision to proceed.

Analysis

In the previous tests, when the weight is set at 0, the results remain at 50 regardless of the defuzzification, implication and aggregation methods. Since using weight 0 makes no sense since project cost cannot be the same value across all the scenarios based on the effort, expenses and travel time, it is ignored as it was in the previous tests. However, in the results from the remaining weights across all defuzzification methods, the differences were very subtle.

When noticing, except for weight 0, the lower the weight starting from 1 using the OR operator with the sum aggregation method, the results tend to decrease slightly compared to finding the decrease of values when the weights are lowered. In contrast, while using the centroid and bisector, they intend to increase whereas the values across the weights using LOM defuzzification fluctuate. When compared to the prod implication method, the results are consistent across all the weights. The initial result is set based on the defuzzification method; then when the different weights are applied, the results are the same with no change.

Test Three: Weights against the default Max Aggregation and Min Implication

Changes

In this test, we will take the current system and its results by testing different values for the weight values of 1, 0.75, 0.5, 0.25 and 0 for all the rules across all the defuzzification methods to see if there will be any significant changes to the system and whether or not any will suit the system and provide an appropriate output.

Expected Outcome

Not expecting to get significant changes with the current results, especially for the AND operator, but for the OR operator we expect there will be significant changes. It may help deciding whether or not the change of weight will be beneficial to the system or if other changes need to be made to the system as a whole, in particular, changing the functions.

Results of the test

Firstly, the results from using the AND operator across all the weights do not appear to show much change. However, during testing using the OR operator, the results of changing the weight appear to be more significant than the AND operator. The most consistent result was when the weight was set to 0 across all defuzzification methods using both operators; there was no change. In every instance, and the return value was 50%.

Defuzzification

In this test, all five defuzzification methods have been used to test against all the weight changes with the two operators AND and OR. The weights used were 1, 0.75, 0.5, 0.25 and 0.

Analysis

Throughout the changes to the weight for the AND operator across the defuzzification methods, there was little change of any significance to the output based on the expected outputs. However, in some cases, the results from the OR operator across the defuzzification methods appear to show more noticeable results compared to those expected for the output.

Looking at Centroid, these results take the effort of 93%, expenses at 100% and travel time at 100%. I would then expect to have an output of 70–90% because the more effort is required for the project then, the more the project is going to cost, however, in this instance it is giving an output of 50% which is slightly lower when using the OR operator. So, in this case, the Centroid defuzzification method would not be the best suited across all different weights.

The Bisector defuzzification method results in a similar output to the centroid with the expectation too when the project effort is at 5%, expenses at 10% and travel time at 0% giving a result of 19%. Similarly, when the effort is at 93%, expenses at 5% and travel time at 5% resulting in an output of 91% again, this was the expectation to the output result, however, across the other inputs, the values are not consistent and provide the expected result. Again, when using the OR operator, the outputs across all weights are inconsistent since the outputs using the OR operator are the opposite of that of the AND operator.

The Middle of Maximum calculates the most likely result. There was an output which resulted in expected output. While using the OR operator where the effort is at 5%, expenses and travel time at 100% with a weight at 0.25 returns an output of 63.9%, which is an output which is expected. However, for the rest of the outputs it was inconsistent, in particular, when the effort is at 95%, expenses at 20% and travel time at 75%, the output was 20% when expecting a moderately higher output. Similarly, it is the same in the case when using the AND operator.

The Smallest of Maximum as the name suggests returns the smallest value from the output. In this instance, while using the AND operator, there were two instances where the output made no sense. There were when the effort was at 95%, travel time at 75% and expenses at 20%, expect a higher output than 50%. Whereas when the effort is it at 93%, expenses at 5% and travel time at 5%, in expecting the output to be lower than the previous output of 95% effort. However, this is not the case, where the output returned a much larger output, in which it is expected to get when the effort was at 95%, expenses at 20% and travel time at 75%.

The Largest of Maximum is the opposite of the SOM, where it returns the maximum value from the output. While using the AND operator, the effort is at either 5% or 93% across all weights are equal. In this instance, the expectation of having different output values. Since more effort is required, the more the project is going to cost. Similarly, when using the OR operator, if the effort is at 5%, expenses and travel time are at 100% and the output is at 100% which again is not the output I am expecting for the system.

All in all, the system could still benefit from further development, especially the second inference system. It was manageable in narrowing down of which operator to use, in this case, the OR operator, however, further minor changes need to be made to the rule base, but most importantly to the membership function and its values. Creating an additional custom defuzzification method should be the next step to see if this will alter the results to become closer to those expected.

5 Discussion

Firstly, research was conducted on the topic to get an overview of what has already been found and if it can assist with testing. Secondly, a prototype of the system needs to be constructed on the basis of information gathered from the research work. Lastly, tests were conducted to see the robustness of the prototype. As expected, there were flaws with some positives and negatives. The tests were thorough and assisted as to which areas are needed to be improved and changed in order to improve the quality of the system outputs.

During the early stages of testing, the rule base was minimal. An increase in the number of rules improved the visibility of the scope of the system, but it hindered at the same time because it was much harder to achieve a match of what was expected compared with the actual results. So, the rule base was later reduced in the tests.

The system has full potential in most if not all environments related to cost estimation however. The limitation does not take into consideration the use of further variables, which would be another way to further develop the system. The system can be developed to something far more significant than it currently is; it has much potential with further testing and fine-tuning of the current system. For a future iteration, it would be useful to reconsider the variables used by comparing them to variables used commonly by businesses, for example, to estimate cost. The variables used currently are based on research and from our findings on how cost estimation is conducted.

The action plan is to conduct further background research, along with that already employed. It is not limited to only research papers but also talking to people within the field which can provide up-to-date insight of what they do on a day-to-day basis which can provide valuable information to produce additional variables for the system.

6 Conclusion

The system has evolved bit by bit throughout the work, and there is still much more testing required, specifically for the second FIS. However, based on the research already conducted where other researchers have already had a similar idea, it is close to being able to apply this system in a real-world scenario, whether that is at a personal project or in business planning. The system still requires further testing, as mentioned in the discussion section, but the system has been put through its paces during the testing phases.

The scepticism during the development of the project is the use of the fuzzy logic toolbox. In the essence of that, when prototyping a system, there are two issues which were encountered. The first being that we are dealing with two inference systems combined with the output from the first inference system, which is then fed into the second system as an input. You can only create one system at a time and cannot combine the two systems. The other is that when you have a system which has been prototyped, then the need to convert the code from a .fis file to a .m file in itself can be a slight inconvenience but it just means more time needs to be spent. However, on the contrary, the option to save that time we could potentially write the code directly in the .m file, thus bypassing the toolbox completely but ultimately is down to preference. In this case, we prefer the visual representation to have a clear vision of the direction of the system.

Despite this, the opportunity to develop a system is to realise the idea of a system that handles business planning and estimating project costs. Applying the concept of fuzzy logic does not only give me a better understanding of how project cost estimations are conducted, but how effectively fuzzy logic can assist in getting better results in cost estimations; this is based on the research conducted at the beginning of the paper within the literature review and from the tests already conducted. This proves there is great potential for the use of fuzzy logic in cost estimations in almost all situations where cost estimations are required, and we believe it is not limited to business use; it could also be applied to other situations.

References

1. Zadeh, L. A. (1965). Fuzzy sets. *Information and Control, 8*(3), 338–353.
2. Attarzadeh, I., & Hock Ow, S. (2010). Improving the accuracy of software cost estimation model based on a new fuzzy logic model. *World Applied Sciences Journal, 8*(2), 177–184.
3. Reddy, C. S., & Raju, K. (2009). Improving the accuracy of effort estimation through fuzzy set representation of size. *Journal of Computer Science, 5*(6), 451–455.
4. Bezděk, V. (2014). Using fuzzy logic in business. *Procedia - Social and Behavioral Sciences, 124,* 371–380.
5. Doloi, H. K. (2011). Understanding stakeholders' perspective of cost estimation in project management. *International Journal of Project Management, 29*(5), 622–636.
6. Kharola, A., & Singh, S. (2014). Development of fuzzy logic model for project management success (PMS). *PM World Journal, 3*(1), 1–11.
7. Marandi, A., & Khan, D. (2017). Software quality improvement and cost estimation using fuzzy logic technique. *International Journal of Applied Engineering Research, 12*(16), 5433–5440.
8. Sharma, V., & Verma, H. (2010). Optimized fuzzy logic-based framework for effort estimation in software development. *International Journal of Computer Science, 7*(2), 30–38.

Artificial Intelligence in FPS Games: NPC Difficulty Effects on Gameplay

Adam Hubble, Jack Moorin, and Arjab Singh Khuman

Abstract This report explores the use of fuzzy logic within computer games, with specific respect to their use of Artificial Intelligence (AI) within the games' enemy Non-Player Characters (NPCs), in order to affect the game's overall difficulty. The way in which AI is affected varies across different games; games within the same genre often share multiple statistics and values, and these can be applied to an NPC in order to make the game easier or harder. Games within the First-Person Shooter (FPS) genre, for example, can always affect their difficulty by changing an enemy character's accuracy with weapons or overall damage output as these would all change how likely they are to defeat the player in a combat scenario. In this document, we will be detailing the development and structure of the multiple input Mamdani styled fuzzy inference system (FIS) that we developed in order to rate a given NPC's difficulty based on the rankings they have been given for these shared statistics.

Keywords Artificial intelligence · Non-player characters · First-person shooter · Mamdani · Fuzzy controller

1 Introduction

The idea and use of fuzzy logic is particularly useful within the field of Artificial Intelligence, because it allows us to apply human-readable labels to more complex logical problems thereby making them easier to understand and develop even further. It also allows us to create more robust systems as we are dealing with groups of inputs instead of precise values.

We decided to utilise fuzzy logic for the development of this application, both because of its popularity [1] within the field of Artificial Intelligence systems and its usefulness in quantifying uncertainty. In our application, we focused specifically on the strength and overall skill of different AI-controlled enemy characters within objective-orientated game modes, in which NPCs can challenge the player with

A. Hubble · J. Moorin · A. S. Khuman (✉)
De Montfort University, Leicester, UK
e-mail: arjab.khuman@dmu.ac.uk

© Springer Nature Switzerland AG 2021
J. Carter et al. (eds.), *Fuzzy Logic*,
https://doi.org/10.1007/978-3-030-66474-9_11

varying levels of difficulty. To ensure overall realism from our calculations, we have also investigated the significance of each characteristic based on the impact it has on the overall difficulty of the enemy character to ensure that we can accurately determine how to provide a balanced and satisfying challenge for players with different skill levels.

To present our data, we created a series of interconnected Mamdani styled fuzzy inference systems [2]. Each system focuses on a different NPC characteristic and has been grouped together with similar characteristics for clarity. We have grouped the sensual ability of NPCs, the objective ability of NPCs as well as the overall lethality of the weapons they are using. Each fuzzy inference system makes use of a list of 'inference rules', stored within each system's rule base, to determine how the given values for each characteristic will affect the overall difficulty of the NPC. This process of using imprecise, or fuzzy, variables in order to reach a conclusion is known as 'Approximate Reasoning' [3]. The design for each system is further detailed and justified in this report, alongside the testing conducted to help improve the accuracy of each systems crisp output values. It is worth noting that we developed this system using the software application MATLAB for the development and testing of our application and implementation of its rule bases and Microsoft Excel for the design and development of the rules. Excel was also used to create and store a data sample that would be input into the system in order to test the accuracy of its rule base.

2 Background and Motivation

The concept of data being governed by logical rules and having a degree of association to sets of data is the focal point of this literature review section. The literature review content will be surrounding the topic of AI within video games, specifically relating to how varying NPC characteristics can present players with different levels of difficulty. This methodology intends to demonstrate real-world game development considerations for AI characteristics and allow for a broad range of potential NPC difficulties in order to cater to the vast potential skill levels of experienced players. It is necessary for successful game development to provide a varying level of NPC difficulty since more skilled players may want to increase the difficulty of a game in order to add more of a challenge, while less skilled, or casual, players may want to make the game less difficult in order to progress through it more easily. It is important to acknowledge the factors of AI within NPCs that create this impact, as well as the logical reasoning behind the rules that are used to determine the characteristics utilised by the NPCs.

The quality and vastness of a game's AI have also become an important selling point for video games [4], with games such as *The Last Guardian* receiving a commendation for its complexity and overall success at, in this case, creating a believable curious animal AI for the game's giant companion character 'Trico' who over time not only can be trained to perform complex commands including jumping onto ledges and destroying large obstacles but also often wanders independently from the player and can potentially decide to ignore the player and take a nap or even start a hissy fit if it hasn't been fed.

It is imperative to outline the fact that the use of fuzzy logic for AI systems within video games is essential for creating a 'dynamic' of NPC behaviours and difficulties, as it pertains to be a method of adding realism to video game worlds by mimicking the way a human would think. An example of this mimicry would be when an NPC has an interaction with one or more objectives (dependant on the game mode), as well as being aware of, and able to combat against, other NPCs and players. What implication does this have on its level of difficulty? How can this difficulty be determined and how can it be measured? These are some of the factors that our fuzzy inference systems must consider, hence why we are going to investigate varying aspects of NPC difficulty, in the proceeding document.

As for our initial source of interest, we wanted to explore the difficulty adjustment of NPCs in video games, specifically within the first-person shooter game genre, as the focus for this report. As we understand, NPC difficulty across video game genres incorporates multiple varying mechanics and game design patterns that are dependent upon the genre a game belongs to. These similarities are typically referred to as 'genre conventions', for which we wanted to gain an understanding of the conventions that make up the AI characteristics of NPCs, within FPS games. The article 'Dynamic Scripting Applied to a First-Person Shooter', which we read to gain a better understanding of our subject matter, explores the development of an AI system for use within FPS games that are fully adapted to show player abilities in any given situation and aim to provide an 'immersive and unpredictable game experience' [5] for players. The article also mentions 'rule-based behaviours' [5] that inform the way in which NPC difficulty can be determined from the use of rules that govern AI characteristics or 'components' [5]. To note, the article predominantly focuses on dynamically changing NPC difficulty by adjusting the rule weighting that governs the AI characteristics following an in-game encounter. However, the difficulty that we configure for NPCs within the fuzzy inference systems (FIS) is static, and therefore means that a player would have to select the difficulty of the NPCs that they want to encounter. With respect to our initial objective of designing and building a fuzzy logic system using one or more FIS, the article initially enables us to acknowledge how NPC difficulty can be determined through the levelling of AI characteristics of an NPC that are aligned by rules. One example of the rules shown within the article demonstrates the relation between the NPC's sensual ability, to detect and see a player character and to engage with them when they're within combat range, which is itself relative to the weapon that the NPC is using.

With continued respect to the weapon types that the article explores, there is a noticeable trade-off made between the hit damage and firing rate of the NPC's weapon. The article refers to this trade-off when describing how a machine gun weapon might shoot 'faster overall but output less damage than perhaps a rocket launcher [would]' [5]. There are also considerations for the '[weapon's] distance from the final point of impact' [5], where there is a decrease in the damage output of a weapon when it is fired over a long distance, thus mimicking the real-world physics concept of 'damage falloff', which refers to when a projectile loses speed and therefore it's resultant impact force, it causes less damage on impact. This concept creates a need for enemy NPC AI to attempt to move towards the player/s so that they can inflict more damage on them. Movement itself is often referred to as one of the 'four distinct entities or components' [5] of a typical FPS game AI. As noted by the authors of the paper, the typical components of an FPS game's AI components 'behaviour, movement, animation and combat' [5], all of which when combined, give the overall difficulty of the controlling NPC. The movement of an NPC is referred to as the combination of speed of the NPC's movement and its navigation through the game environment. With consideration to our FIS, we can adapt movement to refer to the players' mobility and navigation to objective sites. Also, the 'objective' or 'behavioural' [5] AI characteristics of an NPC could determine the NPC's ability to conform to an 'objective orientation of gameplay', whereas the 'animation' and 'combat' [5] AI characteristics could collaboratively determine an NPC's ability to control weapons and respond to player engagement.

For our second source of interest, we primarily wanted to investigate the significance of the weaponry used by the enemy NPC within their overall difficulty in-game. We also wanted to gain an understanding of the effect that a weapon's overall lethality has, as well as further exploring the range of AI characteristics of an NPC that can be restraint or enhanced, depending on the weapon that they have equipped. This focus arises from the vast variety of weapon types that are available within modern FPS games. The article 'Weapon Design Patterns in Shooter Games' examines the different types of conventional weaponry in these shooter games and the ways in which each class of weapon impacts a specific gameplay style, both for NPCs and player/s. Initially, the article discusses how FPS games 'borrow weapon categories' from 'real-world patterns' [6]. However, the article then talks about the differences between the functionality of real-life weapons and those from a video game, specifically referring to how shotguns in video games often have a 'much shorter effectiveness range than [their] real-life counterpart' [6].

Moreover, the article details weaponry variance as 'aspects' which can be 'considered universal among weapon patterns' [6]; these aspects are accounted for as variables of weapon lethality and are described to vary 'between different weapons within a pattern' [6]. Some of the variables listed make considerations for the amount of 'damage a weapon deals with', the 'range of a weapon', how different weapons equipped 'affect[s] the player's movement' and also how the 'continuity of weapon damage' [6] dealt with is presented, all of which contribute to the overall lethality of the weapon. With respect to our FIS, player movement restraints and enhancements are also considerations of the weapons they have equipped. For example, heavier weaponry impacts the player/NPC's speed and mobility.

Furthermore, the article discusses the impact on the user's gameplay style that a weapon can have. One of the given examples of a weapon profile that the article discusses is that of long-ranged 'sniping' [6] weapons, specifically weapons that are designed to fight enemies at a long distance. The authors discuss combating sniping weapons through the influence on players to find 'cover' to avoid being shot. Meanwhile, another example of weapon class that is exemplified within the article is that of a 'close combat weapon' [6], referring specifically to weapons that are designed specifically for use in close-range fights. This encourages the reverse of the defence tactic used when fighting against sniping weapons, as it would be beneficial in this scenario to create distance between the enemy to avoid being hit.

Our final source of interest came from our need to identify how player satisfaction is affected by the scaling of the NPC's AI characteristics. As a game's AI is so significant to its overall success, we wanted to further develop our understanding of how the difficulty of NPC combat can satisfy players of different experience levels. The book 'AI in Computer Games: Generating Interesting Interactive Opponents by the use of Evolutionary Computation' explores player satisfaction in 'predator/prey' video games, specifically focusing on the contribution that 'behaviour and strategy' have on overall NPC difficulty [7]. The book addresses how 'interactive and cooperative characters work to generate more realism to games and satisfaction to the player' [8]. The book also examines NPC 'behaviours', and how they contribute to the 'vast majority of features that make a game interesting' [9]. For the purposes of this report, we will be discussing the contents of the book with specific regard to chapter "City Food in Zimbabwe: The Origins and Evolution" as it presents multiple ideas with regard to 'entertainment metrics in computer games' [9].

In the first chapter of the book, the author describes AI techniques and characteristics that are used to produce characters with 'intelligent capabilities' [8] and infers that the interactivity and cooperation of such characters can enhance game realism and create an increased level of satisfaction for players. The author refers to 'machine learning techniques' [8] that are able to dynamically adjust the NPC difficulty in-game, as opposed to the use of static NPC difficulties and their associated AI characteristics. Moreover, the chapter suggests that there has been a primary focus in game development on the 'graphical representation of game worlds' as opposed to the development of its 'non-player characters' behaviour', and that players who are seeking more 'intelligent opponents and richer interactivity' have found, and created, a surge in the 'increasing popularity of multi-player online games' [8].

Chapter "City Food in Zimbabwe: The Origins and Evolution" of the article discusses the relationship between the 'believability of NPCs and satisfaction of the player' [8]. The chapter describes one criterion of entertainment when a game's NPCs are 'neither too hard or too easy' [9] and introduces the idea that a game is considered interesting when an NPC is able to kill players 'sometimes, but not always' [9]. Another example of NPC AI criteria that the chapter mentions is the NPC's base behavioural state of being 'aggressive rather than static', which 'increases game interest [by] presenting strategic navigation through the game world' [9]. In consideration of our 'Objective Potential' FIS, the concept of providing game interest though NPC behavioural states is adapted as 'objective defensive and offensive pace'. Varying levels of 'aggressiveness' can be configured to cater to the ideal entertainment level based on the player's capability.

3   System Overview

3.1   Design Considerations

For the fuzzy logic system, we have developed multiple Mamdani style Fuzzy Inference Systems which explore and represent different AI characteristics of NPCs within FPS games and the overall difficulty of the NPC they create. All of the sub-FISs present one of the three categories of AI characteristics, in which their outputs are utilised as inputs for the final FIS system and are used to compute an overall NPC difficulty. To note, the output difficulty of NPC that is computed is dependent on the input data that is passed into each of the systems, in which data is passed from an Excel file with the '.xls' extension. Moreover, each sub-FIS has a basis that correlates to the mentioned topics discussed in the literature reviews, which are proven essential to conventional AI characteristics in FPS games [10] and game replayability [11]. Therefore, the systems we configured make considerations for the sensual and objective abilities of NPCs. Another system has been dedicated to the weapon lethality of a weapon that an NPC has equipped. In the proceeding sub-sections of the system overview, we will detail the input and output variables used to accumulate and represent NPC difficulty and provide justification for each of their ranges.

3.2   Fuzzy Inference Sub-system: NPC Sensual Skill

For the 'NPC Sensual Skill' sub-FIS, we configured three input variables that are used to infer the sensual ability of an NPC. The input variables are used as a representation for the raw ability of an NPC, in which 'sensual' provides relevance to the typical senses: 'touch', 'sight' and 'hearing' of NPCs within FPS games. In relation to the sub-FIS, 'touch' is explored in relation to the NPC's ability to control weapons, whereas 'sight' is explored in relation to the NPC's ability to react or respond to detected player stimuli, within a given proximity; player detection is further explored, in relation to the NPC's ability to 'hear' players within said proximity. Meanwhile, the sub-FIS issues one output variable labelled 'NPC Sensual Skill', which aims to represent the combined sensing abilities of an NPC; the crisp output value the sub-FIS computes is then used to form one of the three input variables in the last FIS, 'NPC Difficulty'.

3.3   Variable Justification

In accordance with the table above, the input variable 'Weapon Recoil Patterns and Sight Kick Control' has a range from 0 to 100 and is presented as a percentage; in relation to the intervals, 0 is the minimum possible value and infers 'very poor' weapon control while 100 is the highest possible value and infers 'very good'; the intervals are representatives of NPC weapon accuracy. The range presented enables weapon accuracy to be comprehended accordingly to the number of bullets missed and the number of bullets hit, when an NPC attempts to shoot a player character. This was taken into consideration of the number of bullets that typically reside in the magazine clips of guns in FPS games, whereby the median magazine size I found was 27 bullets, across my study of the weapon specifications within the games: Call of Duty: Black Ops 4 had 26.96 bullets [12], and Rainbow Six Siege had 25.28 bullets [13]; Counter Strike: Global Offensive had 28.91 bullets [14]. An exampling weapon accuracy of an NPC can be '75%'; in relation to the possible intervals, I have justified this value to belonging to the interval or set 'very good'. This is due to an NPC being able to hit 20 out of the possible 27 bullets it may have in its gun's magazine; this infers that a player would die relatively quickly if weapon damage was not factored by range (discussed later).

Moreover, the input variable 'Reaction Time and Responsiveness' ranges from 0.21 to 2 and is measured as a duration of time in seconds. In relation to the intervals, 0.21 is the minimum possible value and indicates 'very fast' while 2 indicates 'very slow'; the intervals in this context are representative of an NPC's response times when detecting any in-game stimuli. This range encapsulates the NPC's responsiveness to the player/s and allows it to be understood comparatively similar to what we would consider a fast-human reaction time and a longer reaction time. This was taken into consideration of human reaction time, wherein we looked at the results given from a number of human reaction time tests regarding simple reaction time (SRT), to explore the 'minimal time [subjects] needed to respond to a stimulus' [15]. From these results, we calculated the average minimal time required within the scope of this study to be 217.9 ms, 0.217 s, a value which was most often present within in the more youthful members of the study, specifically those between 18 and 24 years of age [15]. Therefore, we used this measure from the study to justify our choice of not having the starting range at any value below this 0.21 s boundary, as anything quicker would have made the game's AI so unfairly quick that it reacts faster than any human possibly could. Meanwhile, the range's maximum value of 2 not only is justified as being a 'slow' reaction time with respect to the study, but also caters to players who might perhaps be unexperienced with video games or are relatively new to the FPS game genre.

In terms of the input variable 'Radius of Player Awareness', this variable has a range from 0 to 100 and is measured in a distance of metres. Relating to the range of intervals, 0 is the minimum possible value and infers a 'very close' awareness radius while 100 infers a 'very far' radius. Each interval represents the NPC's proximity for detecting player stimuli, stimuli referring to either the detection of sounds produced by a player character or a visual sighting of a player character, as mentioned previously. The range that we have configured allows for the NPC's proximity of awareness to be considered parallel to that of real-world subjects. To find these values, we conducted another study into the typical awareness radius given to enemy NPC's and from our findings we were able to discover that most game developers attempt to give their AI characters awareness radii that are relative to the real world. Therefore, we arrived at our range on the justification that 'very far' is relative to an awareness radius of 100 m and beyond. With consideration of the size of First-Person Shooter game worlds or maps, 100 in-game metres and beyond is a highly significant amount of map coverage. Meanwhile, 0 represents the lowest possible value for the NPC's range of awareness, since the range cannot be negative within the context of our study and is relative to an NPC's position instead of its facing direction. The interval 'very close' would typically be used by an enemy NPC that is combating players who are considered 'new' to the game and would want an easier time fighting.

With respect to this FIS output variable 'NPC Sensual Skill', the variable ranges from 1 to 5 in the representation of the given NPC's overall sensual ability and is directly dependent on the values given for each of the previously specified input variables. In relation to its given intervals, 1 is the lowest possible value and infers 'very low skill' while 5 is the highest possible value and infers 'very high skill'. We chose this scaling method in order to illustrate each level as a unit, which is why the overall range of 5 correlates to the number of interpreted skill levels available as outputs. We chose 1 to be the starting range value, as we believed that it is better suited to representing the lowest level of sensual ability than 0 as we decided that that would instead logically infer that an NPC has no sensing ability at all.

3.4 Fuzzy Inference Sub-system: NPC Objective Potential

For the 'NPC Objective Potential' sub-Fuzzy Inference System, we have created three input variables that are used to infer the objective ability of an NPC. Each input variable within this sub-FIS is used to compute how well an AI-controlled NPC can present a gameplay style that adopts an objective focus. This specifically caters to the objective-based game modes within FPS games, as referred to in previous sections of this report, in which the objective ability of an NPC is composed of its 'priority to interact with an objective site', the 'pace of gameplay it presents at or around objective sites' as well as its 'responsiveness to new occurrences of objective sites'. Furthermore, the sub-FIS also creates a single output variable titled 'NPC Objective Potential' that aims to represent the combination of objective capabilities that an NPC exercises in-game. This 'crisp' output value the sub-FIS computes is also used as an input in the overall FIS to calculate the NPC's overall skill level.

3.5 Variable Justification

The input variable 'New Objective Responsiveness Time' has a range of 0 to 10 and is a measure of time duration, in seconds. 0 infers a 'very fast' responsive time, while its counterpart 10 infers a 'very slow' reaction time. The intervals in this context are representative of an NPC's responsiveness to new occurrences of objective sites. The range given allows the NPC responsiveness to be understood in relation to what we as humans consider fast or prolonged acknowledgement. The speed at which an NPC acknowledges the location of a new objective site accounts for how quickly an NPC responds to navigating to the site. However, for an NPC to exercise a quickened navigation to an objective site, the NPC must demonstrate a 'high priority' for objective interaction. With respect to our intervals, the minimum responsive time of 0 accounts for an instantaneous acknowledgement, meanwhile the maximum possible value of 10 infers a 'very slow' response time, and is typically used when playing against 'newer' players who are perhaps less knowledgeable of objective site whereabouts and the type of interaction needed at an objective site in order to score points.

The next input variable 'Objective Defensive and Offensive Pace' can range from 0 to 100. This variable is used to indicate the 'aggressiveness' of an NPC, in relation to defending and offending objective sites. In relation to the intervals, 0 infers a 'passive' NPC while 100 infers an 'aggressive' NPC. This value considers both the 'behaviour' and 'movement' AI components of an NPC in a First-Person Shooter game [5] and determines whether an NPC is either effective or ineffective at maintaining a constant interaction with objectives sites. We believe that the use of a percentage here is best for measuring this type of value as we could consider this functional 'aggression' when dealing with enemy NPCs or players to be a position on a spectrum, however, since there is no clearly defined numerical range for this spectrum, we can use a percentage to accurately show how proportional the NPC's aggression is as a portion of its maximum possible state, for which the minimum value 0 correlates to a minimal capability for being able to maintain and compete for objective site control and is better suited to 'newer' players, while conversely the maximum value of 100 would enable a constant sense of challenge for players, with NPCs showcasing 'aggressive' paces and continuous attempts to pressure players for objective site control. This gameplay style would be tailored to experienced players looking for a challenge. It is important to note that the level of objective site interaction is largely impacted by objective priority level, as previously stated.

In continuation to objective priority, the input variable 'Objective Priority Level' has a given range from 1 to 5. This represents an NPC's focus on the objective orientation of gameplay and measures the priority level that it places upon playing the objective. With respect to the given intervals, 1 infers a 'very low priority' setting while 5 infers a 'very high priority' setting. Each interval in this context represents the level at which an NPC considers interacting with objective sites. The given range encapsulates NPC interaction with objective sites. The minimum range value of 1 dictates that an NPC has a minimal to no objective interaction in an objective-based game mode, whereas the maximum range value of 5 indicates that an NPC will show a continuing attendance to objective sites. Allowing the NPC's 'objective priority level' to change in this way allows for 'newer' players to have an opportunity to learn and adapt to objective focuses [8] in game modes, while experienced players are able to experience more of a challenge when playing.

The output variable of this FIS is 'NPC Objective Potential' and has a range from 0 to 100 in the representation of the NPC's overall objective potential. Its value is dependent on the inputs given for the three previously discussed input variables. We have given this output interval of 0, which infers a 'very low [objective] potential' and 100 which infers a 'very high [objective] potential'. We chose to measure this value using percentages since there is no clearly defined metric for a given 'Objective Potential', however, we can use a percentage value to represent the given NPC's overall success at playing objective-based game modes as a proportion of its total possible objective ability. The maximum range value of 100 correlates to an NPC who is completely focused on the game mode's objective and would therefore provide a high amount of challenge.

3.6   Fuzzy Inference Sub-system: NPC Weapon Lethality

For the 'NPC Weapon Lethality' sub-FIS, we have again configured three inputs variables; these are used to infer the weapon lethality of the weapon/s that an NPC currently has equipped. The overall lethality of a weapon can be computed from each input variable value passed into the sub-FIS as a specification and is used to give an indication of how able an NPC is to kill the player/s. A weapons lethality is factored by 'weapon damage falloff', 'weapon fire rate' and 'weapon mobility'. These are conventional gun mechanics within FPS games and differ between each weapon class and weapon individually [6]. Furthermore, the sub-FIS computes the overall 'NPC Weapon Lethality' as an output variable which itself is then fed into the 'NPC Difficulty' FIS as the third and final input.

3.7   Variable Justification

With respect to the first input variable, 'Weapon Damage Falloff', we have given it a range from 0 and 100 m. For the intervals, 0 infers a 'very short range' while 100 infers a 'very long range'. Each interval represents the range from the bullet's initial firing point at which the projectile stops decreasing in damage or plateaus. The array of potential values that we have given for this input allows for the damage falloff to be considered in parallel to real-world environments and concepts. In consideration of game worlds, 100 m is a large distance for a projectile to travel and it is therefore possible that an in-game bullet could travel further than 100 m in the game world; it is likely that the damage falloff will plateau at or before this distance is reached. The minimum possible value is set to 0, so that our system can consider melee weapons, which have no damage falloff as they do not emit projectiles.

The next input variable 'Weapon Fire Rate' ranges from 0 to 2000 and represents the number of bullets that a given weapon can project in a minute. This is referred to in FPS games as 'Rounds Per Minute' [16], or RPM for short. We chose to measure the fire rate of a weapon in minutes as even though many weapons in FPS games present a slow fire rate, it is reasonable to assume that they will be able to fire a bullet at least once within a minute. In relation to the intervals, a minimum input of

0 infers a 'very slow [weapon fire] rate' while the maximum input of 2000 infers a 'very fast [weapon fire] rate'. The maximum possible value for this input is so high as since we are measuring the amount of times a weapon can fire within one whole minute, it is entirely possible for quick-firing weapons, like machine guns, to fire an incredibly high number of rounds within such short period of time. Thankfully we can easily convert the given RPM value into Rounds Per Second, by dividing the value by sixty; this is useful for comprehending the firing rates of slowly projecting weapons. We decided to make the minimum possible value for this input 0, so that the melee weapon category can be considered, in line with melee weapons not emitting projectiles. With respect to gameplay, the faster a weapon can be fired, the more potentially lethal it is.

For the final input of 'Weapon Mobility', we gave it a range from 0 to 100%. We chose to use percentages as the unit of measurement for this variable since it would be impractical for someone to measure the speed at which they can manoeuvre a weapon, so we decided to give the weapon a mobility percentage to present the weapons mobility on a scale from 0, inferring a weapon that is completely immobile, to 100, which infers a weapon that doesn't have any hindrance on mobility at all. For the intervals, 0 infers a 'very slow [mobility] rate' while 100 infers a 'very fast rate'. This input variable allows for considerations to be made for any, mountable or typically heavy weapons that are less manoeuvrable than others.

The overall output variable of this system, 'NPC Weapon Lethality', has a range from 0 to 100%. Similar to the 'weapon mobility' input, we chose to measure the variable as a percentage of its total possible value lethality since there is no obvious suitable measurement for describing the lethality of a weapon. With respect to the intervals, 0 infers that the weapon has a 'very low [overall] lethality', an overall lethality rating of 0 specifically would infer that the weapon is completely harmless, while the maximum possible value of 100 infers that the weapon has a 'very high [overall] lethality'. The total lethality of an NPC's weapon can be considered to be directly proportional to the NPC's overall difficulty, as the more lethal their weapon is, the faster they will be able to kill the player/s.

3.8 Fuzzy Inference Sub-system: NPC Difficulty

For the overall FIS, 'NPC Difficulty', there are three inputs, each of which is the output from one of the previously discussed sub-FISs. This allows our FIS to ultimately consider nine different input variables in a much simpler format. In order to calculate the overall NPC difficulty, the FIS must read-in the computed output values from each of the sub-FISs which are stored within a single '.xls' extension file for convenience. As previously stated, the FIS computes NPC difficulty relative to the NPC's 'sensual ability', 'objective potential' and its equipped weapon's 'weapon lethality'.

3.9   Variable Justification

We decided to give the FISs, 'NPC Difficulty' output, a range from 0 to 100%, to represent the NPC's difficulty as a proportion of the highest overall possible difficulty. Relating to the output given intervals, a difficulty rating of 0 infers that the NPC has a difficulty of 'new', while its counterpart of 100 infers that the NPC is operating at 'veteran' difficulty. The large range of possible NPC difficulties is so that players of all kinds of ability can play and enjoy First-Person Shooter games.

4   Experimental Design and Evaluation

4.1   Initial Fuzzy System Design: MATLAB

The entire system was developed and tested using MATLAB. When developing the system, we made use of three different types of membership functions, or MFs, to determine the degree of membership to which each FIS's input value has to a given set. These functions are triangular membership, a function which shows the gradual linear progression of membership from zero to one and then back to zero forming the shape of a triangle; this is ideal for sets with one value at peak membership. Trapezoidal membership functions, which are similar to triangular membership, not only show the gradual progression of the membership degree from zero to one but also present a plateau region before returning to zero; this is ideal for sets with a group of values with peak membership. Lastly, Gaussian membership functions which are also similar to triangular functions show a membership degree gradually ascending from zero to one and back again, however, the gradient of the line also changes and creates a curved line; this is ideal for sets where the membership degree to a set varies exponentially.

4.2   System Functionality Testing

In order to ensure that each of the systems functioned as intended, we tested the system using a range of different inputs to check that the system computes the desired output. To implement these tests, we created a spreadsheet file that contained multiple sets of input data that was fed into the system. Once the system had computed the input data, it wrote and therefore stored the outputs for the three sub-FISs and overall 'NPC Difficulty' output into the spreadsheet as well. All the data that was both read-in from and wrote-out to the test data spreadsheet was also output to MATLABS's command window (if successful), during the system's compilation process. Fortunately, the system did not present any errors as it managed to both compile correctly and output

Table 1 System functionality test, data samples for writing to file

'NPC Sensual Skill' input variables			'NPC Objective Potential' input variables			'NPC Weapon Lethality' input variables		
Recoil and kick management (%)	Reaction and responsiveness time (s)	Radius of environmental awareness (m)	New objective responsiveness (s)	Objective defensive and offensive pace (%)	Objective priority level (level)	Weapon damage falloff (m)	Weapon fire rate (RPM)	Weapon mobility level (%)
60	7	87	4	56	5	7	755	90
17	4	30	2.1	5	2	26	1200	40
72	1	54	1	88	4	79	450	74

Table 2 Test data, used to determine each input variable value within every sub-FIS. This data is located within the file 'MultipleNPCSkills'

Reaction and responsiveness time (s)	Radius of environmental awareness (m)	New objective responsiveness (s)	Objective defensive and offensive pace (%)	Objective priority level (level)	Weapon damage falloff (m)	Weapon fire rate (RPM)	Weapon mobility level (%)	Expected difficulty (skill level)
2.5	15	10	14	1	10	180	20	New
1.4	47	3.3	37	2	20	350	40	Novice
1.9	51	2.25	54	3	50	600	50	Regular
0.21	100	0	100	5	100	2000	100	Veteran
2	26	4.1	32	2	20	500	50	Regular
1.3	45	2.5	65	3	50	600	45	Regular
1	41	1.75	57	4	40	675	30	Hardened
0.5	75	1	83	4	60	900	40	Veteran
0.9	60	1.5	70	4	50	600	30	Hardened
1.8	30	5.2	40	1	15	400	35	Novice
2	12	8.4	20	2	15	300	15	New
0.85	20	3	50	1	100	40	40	New
1.5	41	4	62	4	15	1100	50	Regular
0.7	62	0.65	74	4	30	1400	90	Veteran
0.65	84	1.25	86	4	80	450	35	Hardened
1.9	33	2.8	47	3	70	550	20	Novice
1.2	42	1.65	74	2	65	700	40	Hardened
2.2	7	7	51	1	15	30	70	New
1.95	30	2.75	60	3	40	480	30	Novice

(continued)

Table 2 (continued)

Reaction and responsiveness time (s)	Radius of environmental awareness (m)	New objective responsiveness (s)	Objective defensive and offensive pace (%)	Objective priority level (level)	Weapon damage falloff (m)	Weapon fire rate (RPM)	Weapon mobility level (%)	Expected difficulty (skill level)
1	50	1	70	5	70	540	35	Hardened
2.45	18	6.2	9	1	25	660	20	New
1.85	25	7.4	22	1	10	300	30	New
0.4	95	1.1	91	4	60	1100	30	Veteran
1.65	43	2.45	60	4	40	375	50	Hardened
1.7	60	4.7	40	3	30	450	60	Novice
1.55	65	1.9	55	4	50	500	30	Regular
0.45	71	0.6	78	5	50	750	70	Veteran
2.35	37	4.5	30	3	35	600	20	Novice
2.1	20	9.2	16	2	90	60	10	New
0.7	86	0.6	72	5	20	1350	80	Hardened

the results that we had been expecting. In Tables 1 and 2 below, you can see the test data we used to test reading and writing data to the file.

4.3   Rule Base Adjustments

In the code for our system, we had defined a series of rules which define what the output of the system will be when certain inputs are given. The original rule base that we created gave an output for each possible combination of inputs. When we first tested the system with this rule base however, it became apparent that this approach had a considerable impact on the time needed both to compile the system code and to execute it. Fortunately, while writing the code, we noticed that there was a considerable amount of repetition within the rule base, from which we could create more abstract rules that would be able to handle the output for more than one combination of inputs. The first obvious occurrence of repetition in the rule base was due to our decision to make the 'NPC Difficulty' output always be 'very low skill' whenever the 'NPC's weapon control' input is considered to be 'very poor', as we believed that any NPC who struggles to properly use their weapon should be considered to be ultimately non-threatening to the player, as they can barely shoot them. The next occurrence of repetition was within the 'NPC Objective Potential' sub-FIS, where we justified that any NPC who has an 'Objective Priority' of 'very low' has an overall 'Objective Potential' that is also 'very low', as an NPC would have little to no objective site interaction and be useless in terms of scoring points [17]. By identifying these cases of repetition within the rule base, we were able to considerably decrease the total number of rules by replacing the superfluous rules with two abstract ones. After replacing the rules, the overall efficiency of the system was noticeably improved and we were able to reduce the total number of rules in effect, from 500 to 433.

4.4   Membership Function Testing

In consideration of membership functions and each fuzzy system, we have incorporated a number of varying membership functions in order to best suit the representation and membership association of data to a given set, as multiple data sets handle variables with differing units of measure and range. With regard to the 'objective potential' sub-FIS, both the 'objective pace' and 'objective potential' variables were adjusted to use a combination of trapezoidal and triangular membership functions which were better suited for representing constant change and gradual distribution. The alternation to using these types of membership functions meant that the Gaussian membership function was completely removed from the sub-FIS, as it did not provide a sensible degree of membership to each bound set. Similarly, we also changed the membership function for the 'weapon lethality' sub-FIS, 'weapon damage falloff' and 'weapon lethality' variables. 'Weapon lethality' was recalibrated for the use of trapezoidal and triangular MFs, as we believed that the normal distribution of membership created by the Gaussian membership function was inappropriate for this. With regard to 'weapon damage falloff', the trapezoidal MFs were replaced with triangular MFs, at the mid-intervals of the variable. This adaptation enabled weapon damage falloff to be represented and calculated more accurately, as a factor of gradual distance rather than regional distance where it was previously represented by a series of plateaus. Furthermore, 'damage falloff' now adheres to the require-ment of damage being relative to the distance from the impact [5]. Lastly, in reference to the 'NPC difficulty' FIS, the 'objective potential', 'weapon lethality' and 'diffi-culty' variables have had their membership functions altered since the initial system design. For both 'objective potential' and 'weapon lethality' variables, the Gaussian MF was once again removed, for the misrepresentation of the membership asso-ciation changing exponentially. We had opted again for the use of trapezoidal and triangular membership, given their previous success for yielding increased accuracy and more so for enabling the set boundaries to be aligned in intervals of '20%'; this allows for more regular distribution, as we desired. Meanwhile, in a continuance of normal distribution, for computing NPC difficulty, we reconfigured the set ranges for each data set to satisfy this distribution, for which each difficulty of NPC is also partitioned into intervals of '20%'. This was more beneficial in comparison to the previous configuration, which discouraged difficulty balancing; this was shown from the sets 'new' and 'veteran', having a significantly smaller share of distribution as opposed to the other difficulties. In this context, the normal distribution has consid-ered the requirement for player satisfaction, in all skill and experience levels [18]. Therefore, we can justify this alteration to be better suited for suitably determining NPC difficulty.

4.5 Defuzzification Method Testing

Defuzzification is the process of finding a 'domain value' that acts as a representative of a fuzzy set. Similarly, to find an average value from a set of data, there are multiple different defuzzification methods we can use. We need to choose a defuzzification method for each of the four fuzzy inference systems within our program. There are a total of five possible defuzzification methods that can be used: Mean of Maximum (MOM), Largest of Maximum (LOM), Smallest of Maximum (SOM), Centroid and Bisector. We used the Centroid defuzzification method during the program's initial setup, but quickly realised that we would need to more carefully consider which method we used, upon noticing that the results we computed did not align with our expectations. Proceeding onwards, we began to experiment with the use of other defuzzification methods and noted the following.

The initial defuzzification method we tested was Bisector, a method which chooses the point that equally splits the area underneath the line. We had first considered using this method as we believed it was the simplest defuzzification method and would therefore be the fastest to calculate from. This proved to be largely ineffective however, as the program appeared to run at a fairly constant speed regardless of the defuzzification method used. We further noticed that the point does not always reside in the centre of a set, if a given set applies a non-symmetrical membership function, which is a result that we would have preferred.

The second defuzzification method that we tested was the Centroid method. We decided to test this method next as the defuzzification method that we tested previously did not nominate the central point of the area underneath the curve, while this is exactly what Centroid does. Even though this method did return the middle value however, we did decide that we didn't want to use this defuzzification method for every fuzzy set, after we realised that since it returned the central value of every set, it was not an accurate representation of the gradual changes in membership on either side of the function. It is worth noting that this defuzzification method was applied during the initial testing phase of the systems development.

The next defuzzification method that we chose to test was Mean of Maximum, or MOM, a method which returns the mean of the values that share the highest membership degree. We chose to test this method as we decided that getting the average of the values with the highest membership would result in a 'domain value'; whilst it wasn't always a representative of the middle of each fuzzy set's membership function, it would still return the centre of the set's 'crisp' values, being the most significant values within that set and would therefore be a suitable defuzzification approach for achieving the most plausible result.

The next defuzzification method that we tested was Largest of Maximum, or LOM, a method which returns the largest value to have the highest given membership degree. We elected to test this method for one of our sub-FISs, where we wanted to experiment with adding a positive bias onto the systems that were applying the Mean of Maximum method. Whilst we ultimately decided that this was inappropriate for use in the majority of our systems, we believed that it would be useful for our 'Weapon Lethality' sub-FIS, since any available weapon in an FPS game must have an overall lethality for which we could indicate by only considering the upper bound of the weapon's possible lethality.

The final defuzzification method that we tested was Smallest of Maximum, or SOM, a method which inversely returns the smallest value to have the highest given membership degree in comparison to LOM. We considered using this method for the final FIS initially, as it was thought that we would achieve a fairer yield of overall NPC difficulty by enabling an NPC to present some challenge to any given skill level of the player. However, from testing the defuzzification method we discovered that the difficulty ranking assigned to most NPCs was in fact unfair. This was evident as NPCs became exceedingly challenging for their supposed difficulty, which was apparent because of the selection of the smallest value with the highest degree of membership to a set; this enabled NPCs with more advanced behaviours to reside in lower difficulty rankings, which did not correlate with our intentions and expectations for implementing the method, the overall result of our testing being the total abandonment of the defuzzification method from the system.

In the below tables, you can see the data from our tests with the different defuzzi-fication methods on each of the FISs. Cells with values shaded in green show an expected and acceptable result while conversely, cells with values shaded in red show an unexpected result (Tables 3, 4, 5 and 6).

Table 3 'NPC Sensual Skill' sub-FIS crisp output values, exploring each defuzzification method
NPC Sensual Skill sub-FIS: defuzzification method comparison

Test number	Centroid	Bisector	MOM	SOM	LOM
			NPC Sensual Skill: crisp output values		
1	1.32	1.28	1	1	1
2	2.343737495	2.24	2	1.52	2.48
3	3	3	3	3	3
4	4.68	4.72	5	5	5
5	2.547668754	2.6	3	2.52	3.48
6	3.5	3.52	3.47804878	2.6	4.36
7	3.5	3.48	3.5	2.52	4.48
8	4.621621622	4.64	4.76	4.52	5
9	4	4	4	3.52	4.48
10	3	3	3	2.52	3.48
11	1.32	1.28	1	1	1
12	1.32	1.28	1	1	1
13	3.468185389	3.4	3	2.76	3.24
14	4.116940092	4.12	4	3.52	4.48
15	4	4	4	3.76	4.24
16	2	2	2	2	2
17	3.5	3.48	3.5	2.52	4.48
18	1.32	1.28	1	1	1
19	2.343737495	2.24	2	1.52	2.48
20	3.6668	3.8	4	4	4
21	1.32	1.28	1	1	1
22	1.32	1.28	1	1	1
23	4.68	4.72	5	5	5
24	3	3	3	2.68	3.32
25	1.951373855	1.96	2	1.72	2.28
26	3	3	3	2.52	3.48
27	4.226093255	4.32	4.88	4.76	5
28	2.452343339	?.4	2	1.32	2.68
29	1.32	1.28	1	1	1
30	4	4	4	3.52	4.48

KEY:
- Expected: [GREEN]
- Not expected: [RED]

Summary:
- Centroid: (**28 / 30**) expected outcomes
- Bisector: (**26 / 30**) expected outcomes
- MOM: (**30 / 30**) expected outcomes
- SOM: (**28 / 30**) expected outcomes
- LOM: (**28 / 30**) expected outcomes

Table 4 'NPC Objective Potential' sub-FIS crisp output values, exploring each defuzzification method

NPC Objective Potential sub-FIS: defuzzification method comparison

	NPC Objective Potential: crisp output values				
Test number	*Centroid*	*Bisector*	*MOM*	*SOM*	*LOM*
1	10.56097561	10	5	0	10
2	30	30	30	23	37
3	50	50	50	46	54
4	89.43902439	90	95	90	100
5	28.64200477	29	30	28	32
6	50	50	50	45	55
7	57.55600815	55	50	43	57
8	80.55530973	83	91.5	83	100
9	70	70	70	70	70
10	10.56097561	10	5	0	10
11	12.41176471	12	10	0	20
12	10.56097561	10	5	0	10
13	50	50	50	42	58
14	88.77777778	89	93	86	100
15	83.90909091	87	93	86	100
16	38.38709677	37	30	22	38
17	50	50	50	46	54
18	10.56097561	10	5	0	10
19	40	40	40	20	60
20	89.43902439	90	95	90	100
21	10.56097561	10	5	0	10
22	10.56097561	10	5	0	10
23	88.77777778	89	93	86	100
24	60	60	60	40	80
25	30	30	30	20	40
26	55.78947368	53	50	45	55
27	88.00680272	88	91	82	100
28	30	30	30	20	40
29	11.59574468	11	8	0	16
30	89.12403101	90	94	88	100

KEY:
- Expected: [GREEN]
- Not expected: [RED]

Summary:
- Centroid: (**30 / 30**) expected outcomes
- Bisector: (**30 / 30**) expected outcomes
- MOM: (**30 / 30**) expected outcomes
- SOM: (**28 / 30**) expected outcomes
- LOM: (**25 / 30**) expected outcomes

5 Nominated Defuzzification Methods

NPC Sensual Skill Sub-Fuzzy Inference System—Mean of Maximum: For the 'Sensual Skill' sub-FIS, we selected the use of MOM at the defuzzification method, as it allows the skill level output to reach the maximum skill level 5. This is possible

Table 5 'NPC Weapon Lethality' sub-FIS crisp output values, exploring each defuzzification method

NPC Weapon Lethality sub-FIS: defuzzification method comparison

NPC Weapon Lethality: crisp output values					
Test number	Centroid	Bisector	MOM	SOM	LOM
1	10.56097561	10	5	0	10
2	30	30	30	30	30
3	50	50	50	40	60
4	89.43902439	90	95	90	100
5	40	40	40	20	60
6	50	50	50	40	60
7	56.875	55	50	40	60
8	70	70	70	64	76
9	50	50	50	40	60
10	25.73972603	27	30	20	40
11	22.34065934	22	20	0	40
12	50	50	50	50	50
13	40	40	40	20	60
14	87.58823529	88	90	80	100
15	70	70	70	65	75
16	50	50	50	44	56
17	70	70	70	64	76
18	30	30	30	20	40
19	50	50	50	40	60
20	63.96046852	66	70	64	76
21	36.875	35	30	20	40
22	22.34065934	22	20	0	40
23	70	70	70	60	80
24	40	40	40	20	60
25	50	50	50	40	60
26	50	50	50	40	60
27	70	70	70	60	80
28	50	50	50	45	55
29	50	50	50	40	60
30	50	50	50	50	50

KEY:
- Expected: [GREEN]
- Not expected: [RED]

Summary:
- Centroid: (**12 / 30**) expected outcomes
- Bisector: (**12 / 30**) expected outcomes
- MOM: (**12 / 30**) expected outcomes
- SOM: (**12 / 30**) expected outcomes
- LOM: (**30 / 30**) expected outcomes

due to the way it calculates the mean of the maximum antecedent values and thus gives an averaged crisp output, the consequent value. The crisp output categorises sensual skill characteristics within the given skill range of 1 to 5 enabling us to accurately identify the skill level.

Table 6 'NPC Difficulty' FIS crisp output values, exploring each defuzzification method

NPC Difficulty FIS: defuzzification method comparison

NPC Difficulty: crisp output values					
Test number	Centroid	Bisector	MOM	SOM	LOM
1	10.64595035	10	5	0	10
2	30	30	30	30	30
3	50	50	50	40	60
4	89.35404965	90	95	90	100
5	40	40	40	20	60
6	59.8123138	60	50	40	60
7	60	60	60	40	80
8	81.00416687	83	90.5	81	100
9	70	70	70	60	80
10	30	30	30	20	40
11	12.41176471	12	10	0	20
12	10.64595035	10	5	0	10
13	50	50	50	40	60
14	89.25063078	90	94.5	89	100
15	73.63472163	72	70	64	76
16	30	30	30	22	38
17	60	60	60	40	80
18	12.41176471	12	10	0	20
19	30	30	30	20	40
20	74.30337079	73	70	64	76
21	12.41176471	12	10	0	20
22	12.41176471	12	10	0	20
23	87.58823529	88	90	80	100
24	60	60	60	40	80
25	30	30	30	20	40
26	57.5443038	56	50	40	60
27	83.41584158	85	90	80	100
28	30	30	30	25	35
29	12.41176471	12	10	0	20
30	70	70	70	70	70

KEY:
- Expected: **[GREEN]**
- Not expected: **[RED]**

Summary:
- Centroid: (**30 / 30**) expected outcomes
- Bisector: (**29 / 30**) expected outcomes
- MOM: (**30 / 30**) expected outcomes
- SOM: (**26 / 30**) expected outcomes
- LOM: (**13 / 30**) expected outcomes

NPC Objective Potential Sub-Fuzzy Inference System—Centroid: For the 'Objective Potential' sub-FIS, we elected the application of the Centroid defuzzification method, as it allows the objective potential output to reach the maximum possible output range, 'very high potential', however, it does not allow the maximum value to reach its potential maximum value of 100, due to the trapezoidal membership functions we have implemented, in which the maximum degree of membership to the start and end sets (relative to the ranges) typically starts before or after the range boundaries of the variable. But, in comparison to all of the other defuzzification methods available, Centroid provides a more gradual difference in respect of computing crisp output values, which enables a larger yield for variation and accuracy within the data. We briefly considered using the Largest of Maximum membership function for this sub-FIS as it enabled the system's final crisp output to reach the maximum possible value of 100, but the output demonstrated a considerably less gradual difference between the crisp output values.

NPC Weapon lethality Sub-Fuzzy Inference System—Largest of Maximum: For the 'Weapon Lethality' inference system, we nominated to use the LOM defuzzification method, as it allows the system's crisp output value to reach the maximum possible output of 100. We considered LOM to be the ideal method of defuzzification in regard to weapon lethality, as it issues balance between each input variable. Whenever one of the system's given input variable values is considered to be 'very high', the output 'Weapon Lethality' increases dramatically and therefore causes an increase in NPC difficulty also. Our use of the Largest of Maximum ultimately results in an overall fairer yield for 'Weapon Ability', in relation to each NPC skill level, whereas, in comparison to the Mean of Maximum, Smallest of Maximum, Centroid and Bisector methods, 'Weapon Lethality' is considered to be too advanced for the desired NPC skill level. Unlike the use of the Centroid defuzzification method in the 'Objective Potential' sub-FIS, the Largest of Maximum defuzzification method when used in the 'Weapon Lethality' caters to more gradual difference in the crisp output value. This is another reason why we considered it to be a suitable defuzzification method.

NPC Difficulty Fuzzy Inference System—Centroid: For the final fuzzy inference system, 'NPC Difficulty', we had chosen to use Centroid as the defuzzification method, as it offers a higher degree of accuracy when computing the output difficulty value of an NPC, as opposed to the other possible defuzzification methods. Given that this is the final output of the system, we considered accuracy to be the most significant variable in the determination of an NPC's difficulty, as any subtle variation in crisp value could cause the category of NPC's difficulty to change and as such, we chose to apply Centroid. If the abilities of an NPC are not categorised correctly, a player might be matched against an NPC that is either too easy to defeat and thus causes the player to have decreased satisfaction, or too hard to defeat and causes the player to become frustrated with the game and similarly lose interest. As discussed throughout this document, player satisfaction is a hugely important consideration in determining a game's success, for which NPCs should be to balance player performance, in the perspective of being able to kill a player 'sometimes, but not always' [9]. The use of Centroid defuzzification in our final FIS adheres to this requirement and thus is considered a sensible defuzzification method of choosing.

6 Discussion

In the end, our system proved to be able to accurately output the 'Difficulty Rating' of an NPC based on input data given for 'NPC Sensual Skill' ('Weapon Recoil Patterns and Sight Kick Control', 'Reaction Time and Responsiveness' and 'Radius of Player Awareness'), 'NPC Objective Potential' ('New Objective Responsiveness Time', 'Objective Defensive and Offensive Pace' and 'Objective Priority Level') and 'NPC Weapon Lethality' ('Weapon Damage Falloff', 'Weapon Fire Rate' and 'Weapon Mobility'). Initial testing of the system fortunately did not yield any errors, which gave us more time to amend the system, via allowing its rule base to become more robust and by testing all of the possible defuzzification methods. As can be noticed throughout the testing process, the defuzzification methods that we had elected for each system were the better suited and the most considerate methods for each of the system's rule base and for also representing each AI characteristic, in correspondence to the conventional requirements of FPS games [5]. Moreover, from testing each of the defuzzification methods, our design decisions have become more comprehensive which will aid us in the development of future systems or, if necessary, improving the one presented.

If the opportunity came to improve the system, there are a few improvements that we could suggest. Firstly, we could introduce a larger amount of inputs for the system, so that the results computed could be more accurate in the representation of the NPC's overall difficulty. A suggested example for another system input variable could include the NPC's own health supply and if it has any resistances to any type of weaponry or attacks, such as a character carrying a riot shield who is impervious to attacks from their front.

Another concept we could introduce to the system is the complexity of the game's AI. This is especially noteworthy with regards to the NPC's overall difficulty, as in some games the enemies may not have a comprehensive understanding of utilising cover effectively, while in some others they may be able to promote advanced routing strategies such as flanking.

Additionally, we could further improve our system by continuing to review and improve upon the system's rule base. When developing the system, we had initially planned to include many more 'abstract rules' than just the two that we did add. Even though these rules did have a significant impact on the program's overall run time, their absence allowed us to decrease the number of rules in the rule base by 13.4%. If more time was available for analysing our current rule base, or perhaps if we added to it in future, it's possible that more instances of repetition may be noticed and therefore replaced by abstract rules.

In consideration of real-world applications of our system, we believe that it could be used within the video games development industry to determine the difficulty of an NPC in an FPS game, since its architecture is largely based on that of existing first-person shooter artificial intelligence-controlled characters. It is worth noting, however, that recent developments within AI in video games have meant there has been an increase in so-called 'dynamic scripting', which allows for the difficulty of the AI-controlled characters to be controlled dynamically by the game itself. This ultimately means that while an NPC may be configured to have an easier overall difficulty at initial points in a game session, the system may decide to increase their given difficulty if the player/s starts playing better than they had previously. For the current configuration of the system, however, its application would be better suited for use within games that support the more traditional statically selected difficulty.

With final acknowledgements to our testing process, we have accounted for 30 test cases using a broad variety of input data in order to test each data set's highest and lowest possible values, so that we can test for potential erroneous edge cases. While we did find this to be a suitable sample size, if we were to consider further developing this system, or another system, it would be indefinitely beneficial to consider conducting more test cases to ensure our system computes data as accurate and reliable as possible. Also, with regard to the testing process, we would also consider making use of the provided fuzzy logic toolbox in future developments, in order to prototype our systems. This would result in our testing process being significantly faster from being able to use the drag and drop interface, instead of having to code them.

7   Conclusion

In conclusion, the development of this system definitely made for an interesting experience as it combines the study of fuzzy logic and knowledge-based systems with exploring our favourite video game genres and gaining a deeper understanding of their inner workings. Throughout the system's development, we have learnt a significant amount, both about the conventions of FPS games and the different AI components that affect the difficulty of NPCs. Moreover, our understanding of player satisfaction within video games has also increased, as we have learnt more about the design considerations made when developing FPS games. We have developed a comprehensive system that could be used within a real-world environment. As mentioned in the discussion section, future work could include the adding of more inputs to potentially make the system more accurate and to continue to make the rule base more robust by removing any occurrences of repetition, possible by a more exhaustive testing process.

References

1. MathWorks (2019) *What is fuzzy logic?* Available from: https://uk.mathworks.com/help/fuzzy/what-is-fuzzy-logic.html. Retrieved November 21, 2019.
2. Ahmad, Y. A., Riyanarto, S., & Oxsy, G. (2018). Rain detection system for estimate weather level using Mamdani fuzzy inference system. In *2018 International Conference on Information and Communications Technology (ICOIACT)*. https://ieeexplore.ieee.org/document/8350711. Retrieved November 21, 2019.

3. Czabanski, R., Jezewski, M., & Leski, J. (2017). Introduction to fuzzy systems. In *Theory and Applications of Ordered Fuzzy Numbers*. https://www.researchgate.net/publication/320 450324_Introduction_to_Fuzzy_Systems. Retrieved November 21, 2019.
4. Forbus, K. D., & Laird, J. (2002). AI and the entertainment industry. *IEEE Intelligent Systems, 17*(4). https://ieeexplore.ieee.org/document/1024746. Retrieved November 22, 2019.
5. Policarpo, D., Urbano, P., & Loureiro, T. (2010). Dynamic scripting applied to a First-Person Shooter. In: *5th Iberian Conference on Information Systems and Technologies*. https://ieeexp lore.ieee.org/document/5556600. Retrieved November 22, 2019.
6. Giusti, R., Hulett, K., & Whitehead, J. (2012). Weapon design patterns in shooter games. In *Proceedings of the First Workshop on Design Patterns in Games*. https://www.researchgate.net/ publication/261857628_Weapon_design_patterns_in_shooter_games. Retrieved November 22, 2019.
7. Yannakakis, G. N. (2005). Abstract. In J. Hallam & J. Levine. (eds.) *AI in computer games: Generating interesting interactive opponents by the use of evolutionary computation. Edition* (p. 3). University of Edinburgh: College of Science and Engineering. School of Informatics.
8. Yannakakis, G. N. (2005). Chapter 1. In J. Hallam & J. Levine (eds.) *AI in computer games: Generating interesting interactive opponents by the use of evolutionary computation. Edition* (pp. 23–24). University of Edinburgh: College of Science and Engineering. School of Informatics
9. Yannakakis, G. N. (2005). Chapter 2. In J. Hallam, & J Levine (eds.) *AI in computer games: Generating interesting interactive opponents by the use of evolutionary computation. Edition* (pp. 31–40). University of Edinburgh: College of Science and Engineering. School of Informatics.
10. Cășvean, T. P. (2015) An Introduction to Videogame Genre Theory. Understanding Videogame Genre Framework. *Athens Journal of Mass Media and Communications, 2*(1). https://www.semanticscholar.org/paper/An-Introduction-to-Videogame-Genre-Theory.-Genre-C%C4%83%C8%99vean/4dd13db214c63cd19c9d0e97df098170dd75dbc0. Retrieved November 22, 2019.
11. Tan, C. H., Tan, K. C., & Tay, A. (2011). Dynamic game difficulty scaling using adaptive behavior-based AI. *IEEE Transactions on Computational Intelligence and AI in Games, 3*(Issue 4). https://ieeexplore-ieee-org.proxy.library.dmu.ac.uk/document/5783334. Retrieved November 22, 2019.
12. IGN. (2019). *Call of duty: Black Ops 4 wiki guide*. https://uk.ign.com/wikis/call-of-duty-black-ops-4/Weapons_List. Retrieved November 26, 2019.
13. IGN. (2019). *Tom Clancy's rainbow six siege wiki guide*. https://uk.ign.com/wikis/rainbow-six-siege/Weapons_and_Equipment. Retrieved November 26, 2019.
14. Fandom. (2019). *Counter-strike: Global offensive*. https://counterstrike.fandom.com/wiki/Cou nter-Strike:_Global_Offensive. Retrieved November 26, 2019.
15. Woods, D. L., Wyma, J. M., Yund, E. W., Herron, T. J., & Reed, B. (2015). Factors influencing the latency of simple reaction time. *Frontiers in Human Neuroscience*. http://europepmc.org/articles/PMC4374455?fromSearch=singleResult&fromQuery= DOI:10.3389/fnhum.2015.00131. Retrieved November 26, 2019.
16. GameGuideHQ. (2018). *Black Ops 4 multiplayer weapon statistics—Damage, time to kill, rate of fire, reload time*. https://gameguidehq.com/black-ops-4-multiplayer-weapon-statistics-dam age-time-to-kill-rate-of-fire-reload-time/. Retrieved November 27, 2019.
17. Glavin, F. G., & Madden, M. G. (2015). Adaptive shooting for bots in first person shooter games using reinforcement learning. *IEEE Transactions on Computational Intelligence and AI in Games, 7*(Issue 2/June). https://ieeexplore.ieee.org/document/6922494. Retrieved November 23, 2019.
18. Yannakakis, G. N. (2005). Chapter 9. In J. Hallam & J. Levine (eds.) *AI in computer games: Generating interesting interactive opponents by the use of evolutionary computation. Edition* (pp. 205–207). University of Edinburgh: College of Science and Engineering. School of Informatics.

Adaptive Cruise Control Using Fuzzy Logic

Nathan Lloyd and Arjab Singh Khuman

Abstract Modern transportation undoubtedly provides a plethora of beneficial qualities; qualities that not only dramatically improve the efficiency and speed of travel, but also provide materialistic comforts for the inhabitants of the vehicle. Whilst these advancements have generally improved the quality of life for users, it begs the question: can modern technologies be utilized to augment vehicles further? This chapter will engage intelligent transportation systems (ITS), specifically automatic cruise control (ACC) and the utilization of fuzzy inference systems (FIS), analyzing their successful implementation, posing a bespoke system and how the ITS field can be improved further.

Keywords Fuzzy Inference System · Automatic Cruise Control · Mamdani · Intelligent Transportation Systems · Motor Vehicles

1 Introduction

Unlike binary logic, the fuzzy logic paradigm is discerned by its ability to handle overlaps in data due to its modelling of vagueness and uncertainty. Whereas binary logic requires crisp values and fuzzy logic is able to parse imprecise information; thereby increasing the scope of its applicability to real-world scenarios. For this reason, it is the aim of this chapter to investigate the elements influencing a safe drive and to design a system suitable for cruise control; in which there is a wealth of literature. Modern *'smart'* vehicles and the overlapping intelligent transportation systems field are typically outfitted with various assistive technologies such as adaptive cruise control, which is the specific focus of the proposed system. Despite relevant literature, however, the system proposed does not have a relevant data set

N. Lloyd
School of Computer Science and Informatics, De Montfort University, Leicester, UK

A. S. Khuman (✉)
School of Computer Science and Informatics, Institute of Artificial Intelligence (IAI), Leicester, UK
e-mail: arjab.khuman@dmu.ac.uk

© Springer Nature Switzerland AG 2021
J. Carter et al. (eds.), *Fuzzy Logic*,
https://doi.org/10.1007/978-3-030-66474-9_12

to test the system. This is due to adaptive cruise control technologies varying from system to system; each using slightly different parameters. Instead, a common-sense approach to create a theoretical data set would be taken. On which, suitable testing of t-conorm, t-norm and intervals was conducted to ensure the validity of the results and the robustness of the fuzzy system.

2 Literature Review

2.1 Why Does the World Need ITS?

The modern world relies heavily upon transport systems for day-to-day activities: commutes, tourism and delivery services just to name a few. These transport systems are integral to a modern way of living, generally improving the quality of life for those who have access. The key of which is the mobility and independence vehicles provide, permitting access to parts of the world in an instant when compared to travel just a few centuries ago. This immense integration of transport systems into daily life is arguably due to the continuous development of technology and the heavy reliance modern societies place upon it; with there being 38.7 million licensed vehicles within Great Britain alone [8]. The adoption of technology can be recognized underpinning systems from the automotive to aerospace industries. Although these technologies have certainly improved the overall efficiency, safety and wellbeing of consumers, significant literature would suggest that the use of transport systems still poses a significant risk to human health.

Motor vehicles, for example, are known to possess a level of infamy due to their impact on human mortality. In 2018, there were 1,784 reported road deaths in the United Kingdom, with another 25,511 serious injuries reported as road traffic accidents [9]. The trend presented in the statistics is echoed throughout supporting literature in other countries, with the World Health Organization estimating 1.35 million deaths per year; currently declaring it as the eighth leading cause of death [45]. These worrisome statistics of motor vehicles are undoubtedly rooted within its common use by the public, increasing the likelihood of accidents being incurred. Moreover, the likelihood of road traffic accidents occurring within less economically developed countries is extremely high in comparison to EU countries, which have seen a minor net decrease in the number of incidents from 2007–2016 [12]. This decrease is clearly linked to the available technology, infrastructure and regulations employed for road safety,improving post-crash injury and providing enhanced systems for collision prevention. The adoption of better systems clearly holds a substantial impact upon the safety of road users, with its implementation undoubtedly saving lives. The application of intelligent transportation systems not only provides benefits to road users but also provides profound environmental benefits via the improved management of traffic flow.

Whilst issues such as carbon emissions do not have the immediate impact or notoriety of vehicle fatalities, issues arise as a by-product of modern methods of travel, such as greenhouse gases, noise pollution and disease transmission; an issue at the forefront of 2020. An abundance of negative impacts can be attributed to pollution alone, with many governing bodies crediting air pollution from vehicles as a contributing factor for childhood asthma [6, 28]. The cumulative impact of these by-products over time will undoubtedly cause substantial negative impacts on the current public as well as future generations, issues that will further compound as the accessibility of transportation increases. The evident negative impacts current transport systems impose are plentiful and although these could be considered heavily within a philosophical platform, solutions to these problems are now within reach due to the powerful technology now available.

This chapter will prioritize the exploration of motor vehicles as they pose the most significant threat. However, the adoption of intelligent transportation systems amongst other modes of transport is clearly a must for future organizations for the improvement of safety and the minimization of the aforementioned side effects. Naturally, intelligent transportations can be applied within any mode of transport, nonetheless EU regulation has chosen to prioritize the application of ITS towards road transport and subsequent infrastructure [35]. Although planes, trains and boats do not possess the same qualities of a motor vehicle, nor are intelligent transportation systems well established and accepted within their fields, they would greatly benefit from the application of ITS due to their ability to carry a large number of passengers. That being said, as the main focus is upon road vehicles, it is important to establish their inception, history and development until modern vehicles.

2.2 A Brief History of Automobiles and Cruise Control

Since Karl Benz's inception of the modern vehicle in 1885, various technological innovations have ushered in vast improvements in the efficiency, comfort and safety of vehicles. However, there is a continuous view that driving a vehicle is one of the largest risks to human life. There are various elements attributed to this problem which are typically hard to quantify; this specifically makes it a hard problem to solve [11]. This risk to human life is a threat that many accept as part of their day-to-day life due to some misplaced optimism bias, unrealistic optimism and traffic risk perception that are intrinsically linked [7]. World Health Organization estimates that by 2030, it will be the fifth leading cause of death [44]. Often as vehicle policies vary by country, such as driving age, alcohol and vehicle regulatory laws, it is hard to construct a solution that would fit internationally. That said, there are evident key elements that compose a safe journey, and they will be discussed in this paper, as well as make up the foundations of the proposed system to improve traffic safety.

One of the most renowned systems for modern vehicles is Ralph Teetor's cruise control [42]. The system is designed to regulate speed, comfort, simplicity but most of all safety [40]. Traditional cruise control, commercially introduced in the

1960s, allows drivers to maintain a consistent set speed without using the accelerator; regardless of the gradient of the road. Adaptive cruise control, however, takes these features and amplifies them. Instead, ACC automatically adjusts the speed of the vehicle to maintain a safe distance from obstacles; which is what makes it one of the most popular research topics within the intelligent transportation systems field [2]. A practical and common intelligent control system for adaptive cruise control is fuzzy logic, as its simple if–then logic is highly applicable to the action of driving.

2.3   A Brief Introduction to Fuzzy Logic

Fuzzy logic is an alternate method to standard binary logic and set theory. Established by Lotfi A. Zadeh [46], fuzzy logic is capable of measuring data by degrees of truth, rather than needing crisp values. This connotes the paradigm's ability to mimic human decision-making, and so, the application of a fuzzy methodology provides the opportunity to parse vague and imprecise information. Whilst there are no crisp boundaries, there is the capacity for an overlap of sets; individual elements of which are distinguishable by degrees of membership. Fuzzy logic as a tool can thereby enable practitioners to manage the trade-off between complexity and precision when dealing with vagueness and uncertainty; developing inexact models from input–output data [47]. Whilst this concept was originally produced to combine mathematical models and linguistic descriptions from natural language, the paradigm has since seen success in other fields, from data mining to weather forecasting [5, 17].

There are multiple variations of a fuzzy inference system; the following two, however, are the most distinct and commonly used; Mamdani and Tagaki-Sugeno, both of which have their own advantages and disadvantages [16]. The more suitable system for this project for being both the standard system widely supported on MATLAB, as well as having wide interpretability due to its rules consequents, is the Mamdani FIS. A Mamdani style system, as proposed by Mamdani [27], is highly tuned to human inputs and is accepting non-linear variables; classifications or natural statements. This method of control is therefore highly intuitive and can be comprised of highly interpretable rules that allow for a model to develop heuristically.

Mamdani style systems, whilst successful, can often be viewed as controversial when implemented as control systems for safety; especially within vehicles and ACC [32]. A critical issue when these systems are applied to real-world scenarios is the well-known failure to infer from contradictory decisions, an obvious problem when dealing with multiple moving objects [36]. Knowing this, simplifying the system taking only a distance, current speed and goal speed as key elements to manage this issue has been a core method of implementing similar ACC systems [21].

2.4 Better Technology Better systems

Due to the age of previous ACC systems based in this paradigm, many of the constraints are based within the level of technology available at that time. The wide availability and enhancements in modern technology have allowed for vast improvements within the commercial and domestic deployment of improved embedded devices; a topic discussed in Moore's Law and its original paper [30]. These improvements have enabled systems such as Tesla's autopilot, a system that can employ neural networks to train and improve the models for steering and acceleration, and so, the electronics supporting these systems have themselves improved. Tesla's autopilot website suggests a 250 m maximum sensor view [43], a figure which has been core to the development of the proposed system. Further inspiration was gained from the Stop and go system developed by Naranjo et al. [31], in which a braking system was naturally incorporated as one of the main features for a fuzzy ACC system. The key purpose of adaptive cruise control is to modify the speed of a vehicle dependent on the distance of objects directly in front of said vehicle, and, although this system improves road safety, it can be vastly improved by considering other elements.

2.5 Better Inputs: Stronger Outputs

For instance, the impact of hazardous environments when driving can be seen globally from Iran [24] to Canada [29], respectively. This hazardous impact is a problem which is magnified due to the impossibility to create uniform roads over a variety of environments,especially when considering a global approach to this problem. The urban–rural divide as an underlying cause of accidents is often one of the key attributes of a hazardous road. For every 100,000 people, rural counties experience 8 more deaths than urban counties [18]. The location of the road is not the only prevalent factor that increases the likelihood of a motor vehicle accident; the car's ability to sufficiently get traction on the road surface is clearly an element that affects the control of the car. The stability of the road surface can be affected by many factors; weather, attrition and the simple composition of the road can be substantial factors to decide whether a wheel is able to have enough friction with the road. Further environmental factors can be attributed to the weather, fog and mist which are renowned for being incredibly dangerous factors in road traffic accidents as this causes a hindrance to visibility. This environmental factor which would otherwise impair human vision can be mitigated with only specific electronic systems [26], and so, must be an essential part of the system to ensure a similar result.

2.6   The Human Element

As the issue of road safety is vast and encompasses various elements, alternative inputs were considered. The main variable considered was the individuality of each driver; modelling their mental state and driving capabilities [14]. There is a multitude of supporting literature to suggest modelling the performance of a driver as a subsystem [1],as personnel vehicles are manned, the human factor is key. Individual differences between human properties are typically hard to model, especially within a linear fashion, and so, is well suited to the fuzzy paradigm. When modelling the human senses, obvious factors such as touch and vision are considered as they are often viewed as being intrinsically linked with the driving skill [33]. Similarly, cognitive states such as fatigue can be collated together to analyze the performance of a driver; cognitive states typically have catalytic-type effects on the potential of an accident [25]. In parallel, there is supporting literature that models niche elements of a safe journey, such as seat belt use, or compares demographic populations against one another [23]. The latter model utilizes a probabilistic approach and has been defined by linear rules, and so as previously stated, the model will struggle to encapsulate non-linear processes. The implementation of a fuzzy system instead would succeed in capturing non-linear classifications and returning actionable crisp values, thereby being a more applicable system to model on topics such as emotional archetypes. Although the aforementioned systems can be applied to mitigate road traffic accidents, they do so by interacting with the driver which isn't applicable when considering an adaptive cruise control system as it is autonomous.

3   System Overview

3.1   Design Considerations

The proposed adaptive cruise control system has been designed to be implemented recursively to dynamically update the input and output parameters. Taking the distance from the object, current speed, environmental and car quality to inform a live speed adjustment via breaking or acceleration. The recursive element allows the inputs to consistently be recorded, calculating new adjustments on the fly; an important task when many variables are acting upon the drive.

Due to the supporting literature, there is a multitude of factors that have been shown to affect the risk of road traffic accidents, a model all approach was taken to begin with to ensure all elements were considered. Elements include the following:

- The quality of vehicle: (acceleration, braking)
- The quality of driver: (biological vision, reaction speed, emotional state, fatigue)
- The quality of environment: (visibility, stability, road type)
- Current Speed

- Distance from object.

As standard with adaptive cruise control, the current speed and distance from the object were at the core of the system, but as previously stated within the literature review, the quality and performance of a driver clearly have a substantial impact on the likelihood of a motor vehicle collision. Whilst it undoubtedly impacts journey safety, in the context of an adaptive cruise control system, it has no impact on driving and so was not included within the final model. Whilst this is true within this context, due to its importance within a safe journey, it was deemed significant to model as subsystem if the model was to be adapted into a semi-autonomous system. Deciding on how to find the ranges of the remaining parameters and create appropriate memberships was conducted by researching the values that underpin the individual features, and they are discussed in their respective system designs below.

3.2   System Designs

3.2.1   Subsystem: Environment Safety

The environmental quality FIS has three inputs to represent the visibility, quality of road surface as well as the location of the road; rural to central business district (CBD). These three inputs were selected as they had the most impact on collisions and together will return a crisp score used in a later system (Fig. 1 and Table 1).

Justification for variable attributes

Road type, and specifically road location, is one of the most important factors when considering environmental components, with rural locations consistently having higher death rates [3]. The intervals used are based upon the popular city zonation models by E.W. Burgess and Homer Hoyt, respectively [20, 34]. Both models

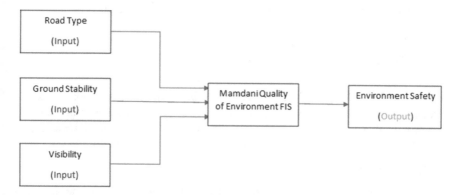

Fig. 1 Visualization of the environmental subsystem FIS

Table 1 Environmental FIS subsystem variables explored

Variable	Type	Range	Memberships
Road type	Input	0–100	Rural, Rural–Urban Fringe, Suburbs, Inner City, CBD and Motorway
Ground stability (%)	Input	0–100	Poor, Average, Excellent
Visibility (M)	Input	0–3000	Dangerously Impaired, Slightly Impaired, Acceptable, Perfect
Environmental safety (%)	Output	0–100	Dangerous, Unsafe, Moderate, Safe, Very Safe

allow for distinguishable ranges of a city centre towards rural areas, which as established is key for measuring the road type. Modelling in this way reveals a clear outlier as motorways are well built up whilst often encompassing a large mass within rural areas. However, a study by the Road Safety Foundation argued that motorways are one of the safest road networks [38], and so, has been included within the CBD.

Naturally, another consideration for this subsystem is the stability of the surface of the road. Car to road friction can be negatively impacted if the composition and material of the road are not suitable; this quality of the road is typically affected by attrition, poor road surfaces, and weather [39]. The range for this variable is therefore a percentage to declare the quality of the road,the higher the percentage, the better the road quality.

The final input for this system is the visibility, whilst there is a technology that can sense distance without using light, an important consideration was the manual override of the ACC in a failsafe scenario. This feature aims to highlight particularly difficult and adverse weather conditions such as dense fog and dust storms, as well as more common hazards such as driving at night. Following the UK highway code, visibility below 100 m is deemed dangerous and requires headlights by law [15], and so, by following this, an appropriate baseline can be found. Whilst the human eye 5 feet above the ground can theoretically see up to 3 miles, research suggests that the human eye can detect a candle flame at 2.6 km away [22], consequently, around 3Km was decided upon for the maximum observable range (Fig. 2).

3.2.2 Subsystem: Car Quality

The car quality subsystem is comprised of two important factors, break and acceleration quality. Theoretically alternative indicators such as mileage could be used; however, many vehicle-based sources presented this and other variables as unreliable indicators of vehicle health, and so, it was not included [4] (Table 2).

Justification for variable attributes

As this system is primarily focussed on the modification of speed, the key two factors are clearly the braking and acceleration systems; with a note to their specific

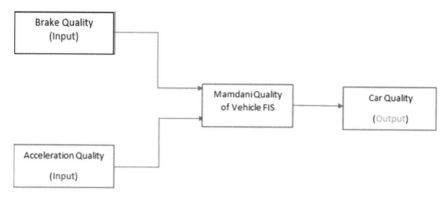

Fig. 2 Visualization of the car quality subsystem FIS

Table 2 Car Quality FIS subsystem variables explored

Variable	Type	Range	Memberships
Brake quality (%)	Input	0–100	Dangerous, Poor, Acceptable, Excellent
Acceleration quality (%)	Input	0–100	Slow, Below Average, Average, Excellent
Car quality (%)	Output	0–100	Low, Below Average, Acceptable, Excellent

quality within the vehicle. Both systems are in a range of 0–100 to represent their quality as a percentage; however, they both have different starting intervals as they are different measurements. Multiple studies have shown brake faults to cause a continually significant impact on accidents, with Gainewe and Masangu's [13] study finding that between 2005–2009 with it contributing an average of 32% of all fatal crashes over the period of study.

Naturally, for changing the speed of a vehicle, the components that affect acceleration must be also considered. Typical research on ACC systems within the fuzzy logic paradigm offers two variations of acceleration; a target and actual acceleration [19]. As an abstraction, however, the quality of the acceleration should mirror the target and actual acceleration scores, with better quality bringing these two figures together (Fig. 3).

3.2.3   Final FIS: Adaptive Cruise Control (ACC) System

The final system for adaptive cruise control calculates a speed adjustment for the vehicle using the two outputs from the former subsystems as well as two new inputs; the speed and distance from the object. These inputs together through the Mamdani FIS allow for an appropriate speed adjustment to be calculated (Table 3).

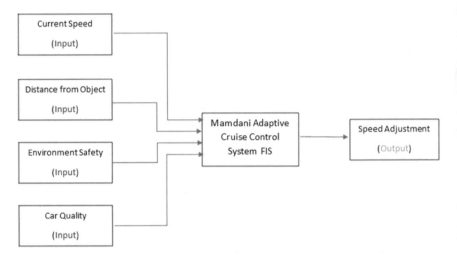

Fig. 3 Visualization of the final cruise control FIS

Table 3 Final ACC FIS variables explored

Variable	Type	Range	Memberships
Current speed (Mph)	Input	0–120	Very Slow, Slow, Moderate, Fast, Very Fast
Distance from object (M)	Input	0–250	Dangerous, Cautious, Safe, Very Safe
Environmental safety (%)	Input	0–100	Dangerous, Unsafe, Moderate, Safe, Very Safe
Car quality (%)	Input	0–100	Low, Below Average, Acceptable, Excellent
Speed adjustment (Mph)	Output	0–120	Very Slow, Slow, Moderate, Fast, Very Fast

Justification for variable attributes

When considering speed as an element on UK roads, 70 mph is the legal speed limit [10]. Many modern vehicles have speedometers that include 120 mph and whilst most never reach this value, accounting for a decline in gradient or a speeding driver before the system is activated, 120 mph was instead chosen as a maximum for the range. Mirroring this speed input into the output simplifies the system, allowing it to become more interpretable when implemented as an adjustment value. The memberships correlate with different types of road: urban to motorways, with the centre of fast being 70 mph.

The final unique input is the distance from an object, the core input for adaptive cruise control. As inferred from the literature review, Tesla's distance sensing technology is the pinnacle of commercial domestic vehicles, and so, an appropriate max range of 250 m can be selected. These inputs together with the inherited outputs from the previous two subsystems are the final elements of the adaptive cruise control system; together evaluating the data to return a speed adjustment.

4 System Testing and Evaluation

The first generation of rule sets for the Environmental Safety, Car Quality and Adaptive Cruise Control was comprised of three variations of membership: Trapezoidal, Triangular and Gaussian. These membership functions provided the ability to utilize appropriate distribution over the fuzzy sets. Whilst membership functions varied between each set, trapezoidal and triangular memberships were typically favoured over entire sets. Often used for the minimum and maximum boundaries of the set, this design choice was to enable the plateau region from trapezoidal to incorporate larger minimum and maximum bounds to provide a greater spread over the entire set; an important factor when these intervals were quite large. Similarly, triangular memberships used in this way were incorporated to provide a wider slope of membership over large intervals as seen within car quality and danger. Gaussian memberships, however, were solely used for middle intervals within the speed and speed adjustment. This design choice was to eliminate select plateau and truncated regions as neither would represent this element realistically.

Originally, the first iteration of the adaptive cruise control system was comprised of 309 rules: 54 environmental, 16 car quality and 239 within the final system. A large rule set can increase the complexity, thereby increasing the computation required to process the inputs. As this is a system that requires a fast response time to ensure a safe output is provided promptly, and this was not adequate. And, whilst each rule was there to map each possible variation of the inputs, optimization was clearly needed and so testing must be conducted to minimize this compute time.

4.1 System Testing

The first stage of testing used random inputs as test cases, a total of 22 tests were conducted to evaluate both the subsystems and final system; noting how the subsystems influenced the final output. This initial testing was formed by using an Excel file to input data into the two subsystems, and these outputs would then be passed into another file and combined with the final system's other two inputs; speed and distance. As the rule base covered every input combination, the final outputs were generally as expected with a few outputs not fitting; this confirmed the applicability of the design chosen, but also presented the need for further testing to improve the system; see Table 4.

As adaptive cruise control technologies vary from system to system, discovering a suitable data set that would fulfil all parameters was impossible, meaning that an artificial data set would need to be created to test the systems. The two subsystems are key factors that are not usually considered within typical adaptive cruise control systems, and so, two of three test scenarios are designed to evaluate the outputs dependent on environment and car quality. Whilst the artificial nature of the data

Table 4 Dummy test case

Input									Expected output	Actual output
Test no	Road type	Ground stability	Visibility	Brake quality	Acceleration	Current speed	Distance	Speed adjustment	Speed adjustment
1	20	80	50	100	100	70	200	Fast	50.02
2	85	80	50	10	50	70	30	Moderate	40.56
3	85	80	50	60	80	100	10	Very Slow	3.93
4	85	80	50	10	10	50	50	Very Slow	27.63
5	85	80	50	10	50	50	50	Moderate	40.56
6	85	80	50	60	80	50	50	Moderate	40.56
7	20	80	50	10	10	50	50	Very Slow	8.15
8	100	80	50	10	50	50	50	Moderate	40.56
9	20	80	50	60	80	50	50	Slow	20.01
10	40	80	50	10	10	50	60	Slow	20.00
11	40	80	50	10	50	50	70	Slow	20.00
12	55	80	50	60	80	50	80	Moderate	39.74
13	60	80	1050	10	10	30	90	Moderate	41.16
14	35	20	1050	10	50	10	200	Slow	20.00
15	25	20	1050	60	80	30	180	Slow	20.00
16	37	20	1050	10	10	10	50	Slow	20.00
17	55	20	1050	10	50	30	70	Moderate	37.43

(continued)

Table 4 (continued)

| Input | | | | | | | | Expected output | Actual output |
Test no	Road type	Ground stability	Visibility	Brake quality	Acceleration	Current speed	Distance	Speed adjustment	Speed adjustment
18	78	20	1050	60	80	40	150	Fast	48.91
19	50	10	3000	10	10	50	50	Moderate	46.77
20	52	70	3000	10	50	50	100	Fast	50.02
21	10	100	3000	60	80	90	50	Very Slow	3.93
22	100	100	3000	85	85	40	140	Fast	69.97

could cause bias within the results, the data set itself has been built using a common-sense approach that displays a clear scenario applicable to the real world. There are two clear benefits of artificial data. The first being the ability to easily predict the outcome of the test and check actual outputs as a comparison. The second being the ability to avoid missing data; therefore, a small but standardized 22 testing examples were used to show a possible flow of time within the scenario provided for each test. As these data sets have been fabricated, multiple tests will be conducted to ensure the reliability of the system, and no human error has made the system impartial.

4.2 Test 1

Each test aims to create a realistic event-type scenario in which the artificial data may replicate true data. In test 1, the vehicle and environment are both in excellent condition, and the main means of testing are the vehicle's distance from another vehicle in the same lane as well as the speed the vehicle is going. This can be seen in Table 5; the data aims to emulate a scenario where there is a low starting speed on a motorway with no direct traffic ahead until the vehicle speeds up. At one point, the vehicle's speed goes above the legal limit, and eventually the car ahead becomes incredibly close; a scenario very likely.

From this test case scenario, the rule base responds to incremental speed and distance changes. The two most notable effects of this system are the reduction in speed when speeding to 70 miles per hour, and then the reduction in speed when the object's distance becomes closer than its own stopping distance [37], as seen in entry 20 and 21. These are the expected results for this scenario. Whilst the results received in this test were as expected, the compute time to receive the results was not acceptable. To ensure all scenarios work effectively, this task of reducing the rule base will be completed after Test 3.

4.3 Test 2

Test 2 continues the established format and is a poor environmental conditions scenario, with all other driving conditions being optimal. This test's design is to view the effect of weather and road conditions on a vehicle. The results of this test are not as expected and by looking at the distribution of the input and output sets, it is clear why. The memberships implemented for stability and the output set are not distributed appropriately; with output unsafe being only to 50% and the input average being at 50%, respectively. A more realistic way of modelling the output membership would be to remove the moderate interval entirely and increase the boundaries of unsafe to cover this area, for ground stability; however, all the intervals will need changing to ensure the average mark is 70%.

Table 5 Speed and distance focus with good environmental and car quality

| Input | | | | | | | | | Expected output | Actual output |
Test no	Road type	Ground stability	Visibility	Brake quality	Acceleration	Current speed	Distance	Speed adjustment	Speed adjustment
1	90	95	2700	95	95	5	180	Very Fast	98.71
2	90	95	2700	95	95	10	200	Moderate	50.08
3	90	95	2700	95	95	15	220	Moderate	50.17
4	90	95	2700	95	95	20	240	Moderate	50.81
5	90	95	2700	95	95	25	250	Moderate	52.18
6	90	95	2700	95	95	30	250	Fast	60.51
7	90	95	2700	95	95	40	250	Fast	69.96
8	90	95	2700	95	95	50	250	Fast	70.00
9	90	95	2700	95	95	60	250	Fast	69.96
10	90	95	2700	95	95	70	250	Fast	70.00
11	90	95	2700	95	95	70	250	Fast	70.00
12	90	95	2700	95	95	80	240	Fast	69.99
13	90	95	2700	95	95	90	220	Fast	70.03
14	90	95	2700	95	95	70	200	Fast	70.00
15	90	95	2700	95	95	60	200	Fast	69.99
16	90	95	2700	95	95	50	180	Fast	70.00
17	90	95	2700	95	95	10	160	Moderate	50.08

(continued)

Table 5 (continued)

Input									Expected output	Actual output
Test no	Road type	Ground stability	Visibility	Brake quality	Acceleration	Current speed	Distance		Speed adjustment	Speed adjustment
18	90	95	2700	95	95	30	160		Fast	60.51
19	90	95	2700	95	95	50	120		Fast	70.00
20	90	95	2700	95	95	60	100		Fast	70.00
21	90	95	2700	95	95	70	80		Moderate	50.03
22	90	95	2700	95	95	70	220		Fast	70.00

4.4 Test 3

Similar to Test 2, Test 3 was designed to test the effects of the subsystem on the final output. The scenario proposed in the data is the rapid reduction in quality of the vehicle whilst all other systems are optimal. This data set aims to mimic car failure due to damage or disrepair; a factor that could inevitably cause a road traffic accident. From testing it, was clear to see that this subsystem produced the worst results within its scenario; as again, the error was due to the lowest membership interval not having a wider spread over the entire set; a fault modelled across each of its features.

4.5 Modifications Based on These Tests

From the three preliminary tests, it was clear to see that the two subsystems were negatively impacting the final output produced by the ACC system; car quality more so than the environmental safety subsystem. To remedy this negative impact, these systems needed their feature membership boundaries modifying.

The environmental safety subsystem needed minimal modification to improve the results of Test 2; the ground stability feature was the main issue within this subsystem; an average score was weighted around 50%. This interval, therefore, needed to be shifted to the right as having an average quality should realistically be 70%; in turn, this required the extension of the unsafe interval. See Fig. 4 for the initial design of ground safety within the environmental subsystem and its transformation into a simpler, yet better model in Fig. 5. Whilst extending the breadth of the unsafe interval over the set, it was apparent that its membership function needed changing to represent a larger maximum membership across a trapezoidal plateau. The output of this system needed similar interval treatment, and this was done by removing the moderate interval and extending the unsafe interval; extending unsafe and safe accordingly.

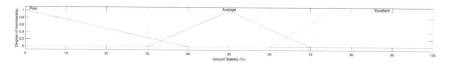

Fig. 4 Original ground stability with poor memberships

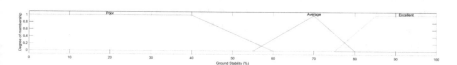

Fig. 5 Updated ground stability with improved memberships

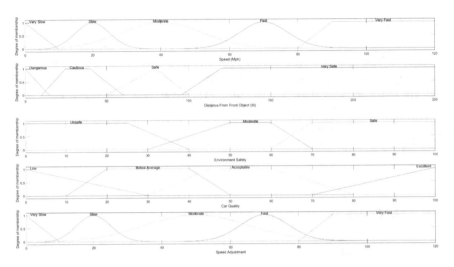

Fig. 6 ACC and subsystem adjustments

Similar treatment was required for the car quality subsystem; however, both the inputs and output needed to be modified. Whilst shifting the interval boundaries, it came apparent that the 'below optimal' is still a region of the dangerous domain, and so, this interval was removed, and the lower bounds extended to fill this region. Like the modifications on the environmental subsystem, the lowest interval was a triangular membership, and this was no longer appropriate as the truncated apex did not accommodate as much data to maximum membership; unlike the trapezoidal membership. A by-product of these modifications was the ability to remove some rules and an output for the car quality subsystem thereby decreasing the rule base for the adaptive cruise control system as a by-product; increasing the speed of computation. These complete modifications for the respective systems can be viewed below; see Fig. 6.

To evaluate the effectiveness of these changes, Test 2 and 3 were rerun and compared. As displayed in Table 6, a clear improvement can be seen in Test case 2, with more moderate scores occurring as expected. The tuning of the memberships was a little heavy handed leading to a small amount of test data being mislabelled from fast to moderate. In the context of this FIS's application to a real-world scenario, driving at a lower speed during middling weather conditions is not a hugely negative impact, and so, for the purposes of this test removing more false fast's over, adding a false moderate is a successful improvement. Test case 3 improvements, on the other hand, were quite marginal, and the lack of meaningful improvements was due to the modification of the set memberships; this tuning although providing marginal gains is still an acceptable result. As previously stated, the modification of the car quality subsystem enabled rules to be pruned from its system as well as the adaptive cruise control system. Overall, the improvements gained from these tests not only

Table 6 Post modification comparison (green: improvements; peach: deterioration)

Test	Test 2			Test 3		
No.	Expect	Original	New	Expected	Original	New
1	Moderate	49.09	49.09	Fast	69.99	60.73
2	Moderate	49.09	49.09	Fast	69.99	70.00
3	Moderate	49.09	50.03	Fast	69.99	70.00
4	Moderate	69.96	50.03	Fast	69.99	69.99
5	Moderate	68.64	50.03	Fast	69.99	69.99
6	Moderate	60.51	50.03	Fast	69.99	69.97
7	Moderate	70.00	50.03	Fast	69.99	70.00
8	Moderate	69.96	50.03	Fast	69.99	69.99
9	Moderate	69.96	50.40	Fast	69.99	50.33
10	Moderate	70.00	50.40	Moderate	69.99	50.32
11	Moderate	70.00	50.40	Moderate	69.99	50.32
12	Moderate	70.00	50.40	Moderate	69.99	50.40
13	Moderate	70.00	50.40	Moderate	69.99	50.40
14	Fast	69.99	50.32	Slow	69.99	50.40
15	Fast	69.99	50.32	Slow	69.99	50.02
16	Fast	70.00	50.33	Slow	69.99	50.02
17	Fast	69.96	69.96	Slow	69.99	50.02
18	Fast	69.99	69.99	Slow	69.99	49.12
19	Fast	69.99	69.99	Slow	63.21	50.02
20	Fast	70.00	70.00	Slow	59.48	50.02
21	Fast	70.00	70.00	Slow	54.63	49.12
22	Fast	60.51	60.51	Slow	50.04	49.12

improved the accuracy of the overall results whilst in three different scenarios but also improved the computation speed by removal of rules, moving from 309 to 262.

4.6 Defuzzification Testing

The initial improvements and testing undertaken until this point have focussed heavily on the inputs and rule base of the systems; however, the final stage of a Mamdani system is defuzzification into a crisp set. This is important to consider as different methods of defuzzification are the key principles for choosing a Mamdani system over a TSK fuzzy system. To modify this final stage, five defuzzification methods were chosen: Centroid, Bisector, Smallest of Maximum (SOM), Middle of Maximum (MOM) and Largest of Maximum (LOM). SOM, MOM and LOM are calculated using the aggregate membership functions, and whilst the adaptive cruise control system only has one interval output that had had a trapezoidal membership function (plateau), the subsystems have many, and so they are unlikely to produce the same value as the maximum is not unique.

Again, when testing for optimal defuzzification methods as the data set is artificial, a precaution of testing twice using two different scenarios was chosen. Now using the improved results from Test 1 and Test 2, both scenarios were run with all five variations of defuzzification. Due to their similar output, solely Test 1s defuzzification will be displayed, although both will be discussed.

First looking at the results of Test 1, the default defuzzification method in MATLAB is centroid and from this, it is possible to see the improvements made from the previous iterations of testing, and data is now as expected when compared to the first results. Comparing these centroid outputs against the bisector presents little deviation between the two, with only miniscule changes to the subsystem inputs, likely having the effect on the final output. These defuzzification methods have returned similar results except for three significant entries: 17, 19 and 21. Whilst the centroid accurately returns the expected result, the bisector has inverted the expected results, with rows 17 and 21 expecting a moderate return, and instead a fast speed adjustment is recommended. Again, with Test 2, a similar result is found with minimal changes between the two defuzzification methods. In contrast, the bisector was slightly more efficient within this scenario as noted with entry 22 being closer to the expected result. Whilst these two alone are very similar, there were slight performance differences between each test, resulting in the centroid being the most optimal of the two; see Table 7.

Table 7 Centroid vs Bisector on Test Case 1

Test no	Centroid			Bisector		
	Environmental	Car quality	ACC	Environmental	Car quality	ACC
1	87.59	94.86	68.39	88.0	95.00	69.60
2	87.59	94.86	50.08	88.0	95.00	50.40
3	87.59	94.86	50.17	88.0	95.00	50.40
4	87.59	94.86	50.82	88.0	95.00	50.40
5	87.59	94.86	52.18	88.0	95.00	51.60
6	87.59	94.86	60.45	88.0	95.00	66.00
7	87.59	94.86	69.96	88.0	95.00	69.60
8	87.59	94.86	69.99	88.0	95.00	69.60
9	87.59	94.86	69.99	88.0	95.00	69.60
10	87.59	94.86	69.99	88.0	95.00	69.60
11	87.59	94.86	69.99	88.0	95.00	69.60
12	87.59	94.86	69.99	88.0	95.00	69.60
13	87.59	94.86	69.99	88.0	95.00	69.60
14	87.59	94.86	69.99	88.0	95.00	69.60
15	87.59	94.86	69.99	88.0	95.00	69.60
16	87.59	94.86	69.99	88.0	95.00	69.60
17	87.59	94.86	50.08	88.0	95.00	69.60
18	87.59	94.86	60.45	88.0	95.00	69.60
19	87.59	94.86	69.99	88.0	95.00	50.40
20	87.59	94.86	70.00	88.0	95.00	66.00
21	87.59	94.86	50.02	88.0	95.00	69.60
22	87.59	94.86	69.99	88.0	95.00	69.60

The final defuzzification methods to test were the maximum bounds, and these produced the most variety within the results over both tests, as shown in Table 8. Initial viewing of the results displays LOM as the worst fit for the systems as it consistently produces results above the max UK speed of 70mph over both Test 1 and Test 2; naturally, this disqualifies the defuzzification method as it breaks the law, and other entries can therefore be disregarded. Alternatively, when implementing SOM over both Test 1 and Test 2, a very low score is returned. Compared to LOM, this is not a negative impact on the final system, when incorporated into a physical system, it may be more suitable due to its very safe results. When looking at this practically within a physical system, using SOM is inevitably a trade-off between travel time and safety which is another issue entirely, and so, centroid remains the best system.

Finally, when looking at the results of Test 1 with defuzzification method MOM, entry 17 becomes misclassified and in turn returns a result lower than expected. In addition, although only a minor amount of speed over, MOM consistently returns 70.20 and whilst this is admissible; when comparing the results achieved by centroid or the overly safe SOM, it is clearly not as effective. When implementing MOM for Test 2, it performs as expected, outperforming SOM in one instance, again overall appearing very similar to centroid and bisector; as expected it is based upon the middle value of the maximum.

From testing of the defuzzification methods, LOM is the poorest method whereas centroid is consistently the best method to use. Bisector and MOM have a relatively high applicability too, but they are limited by some miss classifications that do not affect the centroid. SOM would also be a suitable choice as a replacement for centroid as it returns values only slightly under the expected result; perhaps this would benefit a physical system more if the electronics used to collect inputs were not optimal. Although as mentioned, a limitation of choosing SOM as the defuzzification method is the total journey time, which is a negative factor of the drive and a potential cause for other safety concerns; driving too slow could also cause accidents. Therefore, the centroid is clearly the most optimal defuzzification method to implement within this adaptive cruise control system.

4.7 Aggregation Versus Implication

The specific use of the AND/OR rule operators as well as defuzzification also affects the consequents. The previous iterations of testing have used aggregation within the rule base and so testing the implication is necessary to ensure the current system is the best it can be. The implication operator was utilized for each system. Similar to the previous testing, ensuring no bias due to artificial data was a consideration at the forefront of testing; to mitigate this factor, each test case was used to ensure quality over each scenario when the implication rule was applied; the results of which can be viewed in Table 9.

Table 8 LOM, SOM, MOM test case 1

Test no	MOM			SOM			LOM		
	Environmental	Car quality	ACC	Environmental	Car quality	ACC	Environmental	Car quality	ACC
1	90.00	97.50	70.20	80.00	95.00	20.40	100.00	100.00	78.00
2	90.00	97.50	49.80	80.00	95.00	36.00	100.00	100.00	79.20
3	90.00	97.50	50.40	80.00	95.00	37.20	100.00	100.00	64.80
4	90.00	97.50	50.40	80.00	95.00	36.00	100.00	100.00	50.40
5	90.00	97.50	50.40	80.00	95.00	64.80	100.00	100.00	64.80
6	90.00	97.50	70.20	80.00	95.00	64.80	100.00	100.00	75.60
7	90.00	97.50	69.60	80.00	95.00	64.80	100.00	100.00	69.60
8	90.00	97.50	70.20	80.00	95.00	60.00	100.00	100.00	75.60
9	90.00	97.50	69.60	80.00	95.00	64.80	100.00	100.00	79.20
10	90.00	97.50	69.60	80.00	95.00	64.80	100.00	100.00	69.60
11	90.00	97.50	69.60	80.00	95.00	60.00	100.00	100.00	69.60
12	90.00	97.50	69.60	80.00	95.00	64.80	100.00	100.00	75.60
13	90.00	97.50	69.60	80.00	95.00	64.80	100.00	100.00	79.20
14	90.00	97.50	69.60	80.00	95.00	60.00	100.00	100.00	69.60
15	90.00	97.50	70.20	80.00	95.00	64.80	100.00	100.00	69.60
16	90.00	97.50	49.80	80.00	95.00	20.40	100.00	100.00	79.20
17	90.00	97.50	70.20	80.00	95.00	64.80	100.00	100.00	75.60

(continued)

Table 8 (continued)

Test no	MOM			SOM			LOM		
	Environmental	Car quality	ACC	Environmental	Car quality	ACC	Environmental	Car quality	ACC
18	90.00	97.50	70.20	80.00	95.00	64.80	100.00	100.00	75.60
19	90.00	97.50	70.20	80.00	95.00	64.80	100.00	100.00	75.60
20	90.00	97.50	70.20	80.00	95.00	57.60	100.00	100.00	82.80
21	90.00	97.50	50.40	80.00	95.00	37.20	100.00	100.00	50.40
22	90.00	97.50	69.60	80.00	95.00	64.80	100.00	100.00	69.60

Table 9 Implication results

Test1	Test2	Test3
43.90	43.90	43.90
43.90	43.90	43.90
43.90	43.90	43.90
43.90	43.90	43.90
43.90	43.90	43.90
43.90	43.90	43.90
43.90	43.90	43.90
43.90	43.90	43.90
43.90	43.90	43.90
43.90	43.90	43.90
43.90	43.90	43.90
43.90	43.90	43.90
43.90	43.90	43.90
43.90	43.90	43.90
43.90	43.90	43.90
43.90	43.90	43.90
43.90	43.90	43.90
43.90	43.90	43.90
43.90	43.90	43.90
43.90	43.90	43.90
43.90	43.90	43.90
43.90	43.90	43.90

As displayed within the table, the results were consistently 43.90 over each scenario, clearly suggesting the implication rule does not fit the current composition of the rule base and systems. However, so as to ensure this is not due to the weighting of the rules remaining as 1, further testing will be conducted.

4.8 Implication and Weighting

To ensure the assumption was correct, modification of the weights over each system was conducted, so as to test if the weighting had a meaningful impact on the final output. Modification of the weights is a simple and common way to tune a fuzzy system; the rules weights solely affect the consequents of the system and indicate a measure of importance However, the inclusion of such weightings can in fact reduce the ease of interpretability due to the deviation from simple linguistic expressions. The weighting modification test upon Test 2 was as expected, and again, a repetitive output was returned; see Table 10.

Table 10 Weighting tests

Test no	All tests with a weighting of 0.5
1	42.98
2	42.98
3	42.98
4	42.98
5	42.98
6	42.98
7	42.98
8	42.98
9	42.98
10	42.98
11	42.98
12	42.98
13	42.98
14	42.98
15	42.98
16	42.98
17	42.98
18	42.98
19	42.98
20	42.98
21	42.98
22	42.98

This repeating figure is due to the speed adjustment outputs middle interval of 'moderate' encompassing a large area over the fuzzy set. In addition, the rule base was developed with a one-to-one aggregation in mind, and so, the modification of weights does not improve the system but instead reiterates that the implication rule does not fit the current composition of the system. Whilst this modification does not provide any benefits within the current design of this system, it could perhaps be used if the system was designed differently and not bound by the same constraints. Whilst this clearly reveals the possibility to fine tune the system for varied hazardous scenarios, the system already produces desirable results, and so, rebuilding the system would not be suitable to improve performance.

5   Final System and Discussion

The final system layout inclusive of all positive changes can be viewed in Fig. 6. At this point, no further testing conducted meaningfully improved the overall system and, in most cases, decreased the effectiveness of the FIS.

Modifications through testing have improved the accuracy and decreased the rule base by 15%, calculating an output can still be a relatively exhaustive but a necessary process to ensure the safety of a vehicle. Whilst the computation can be quite intensive when running multiple tests, if this system were to be implemented into a physical system, it would run a single entry at a time, dynamically providing a speed adjustment; perhaps mitigating this computation factor altogether.

Through conducting defuzzification testing, it was clear that the centroid and bisector were very similar and there is minimal advantage to selecting centroid; feasibly with further testing, the bisector could be a more appropriate defuzzification method. Considering SOM is also an option for very safe speed adjustments, and SOM was not selected for the final system due to the possibility of causing traffic or accidents as a consequence of going slower than the speed limit; thereby impeding other drivers. This is an unavoidable problem within this system as there is no way of checking objects to the rear of the vehicle, and this could be rectified if there were additional sensors measuring the distance of vehicles behind; in that instance, SOM would be highly suitable.

There are also limiting factors within the features design, non-most apparent within the environmental subsystem and road type. An assumption of road type gained from research suggests that the more urban an environment, the safer the vehicle; this premise can also be applied to motorways. Evidentially, a city centre and a motorway vary completely when considering max speed, and so, the current system would only work within a real vehicle if it knew the exact road type; if not this could lead to speeding within a CBD. Additionally, car quality is a barebones FIS and extra elements could have been included to improve the output of this system, such as tyre tread and pressure; which themselves could have been another layer of subsystems. A notion which is similar for other elements, that with hindsight could have been modelled differently to improve the reliability of the final output.

Finally, within both literature and the design process, the performance of a driver is an important aspect when considering the likelihood of a road traffic accident. Whilst not important for an adaptive cruise control system, a semi-autonomous system could greatly benefit from the addition of this system. This evaluation of the driver is clearly an important factor throughout intelligent transportation systems, as humans are at the core of the process; often argued to be the leading cause of accidents [41].

6 Conclusion

Typically, research conducted within this paradigm has had a strong focus upon the acceleration of the vehicle dependant on the distance from an object and the current speed. Naturally, these should be the strongest signals within an ACC system, and these two have been important considerations within this FIS. Nonetheless, this approach can overlook external factors such as the environment, car quality and the driver.

The proposed system aims to model these often-overlooked factors as they have a direct impact upon the vehicle, naturally affecting the acceleration and stopping distance. By ensuring to encapsulate these elements within an ACC system, the risk to life can be mitigated, and perhaps, with the correct weighting of the signals and rules, they can be tuned perfectly for hazardous environments. Fuzzy logic is the quintessential tool to model the vague and immeasurable elements whilst on a journey and it is these elements that are modelled within the proposed system.

As a result of testing, the final system developed is much more accurate and efficient than the initial design, and consequently, the development process has been beneficial. The next step would be to test this FIS within a physical or simulated vehicle, so as to ensure the validity of the results gained from the artificial data set, as well as to develop this system further. A simulation could have perhaps increased the ease of testing each element, and this visual representation would have been beneficial to report on. Nevertheless, this basic system could be enhanced and improved upon if necessary and forms a good foundation for improving road safety.

Road safety continues to be an important issue, and as further developments to the ITS field are incorporated globally, the risk to human life is sure to decrease. Until then, road traffic accidents are likely to be an ongoing affliction of travel and until a global approach to the application of ITS systems is undertaken, less economically developed countries are sure to remain the most affected by road collisions. With the technologies supporting travel and the world becoming ever-more connected through these advancements, the opportunity to solve these issues has never been more within reach.

References

1. Baradkar, C., Ganveer, A., Lokhande, S., & Surender, K. (2016). Fuzzy approach for examining the performance of driver. *International Journal of Advanced Research in Computer and Communication Engineering, 5*(7), 663–666.
2. Bursa, M. (2000). Big names invest heavily in advanced ITS technology. *ISATA Mag, 8,* 24–30.
3. Brown, L. H., Khanna, A., & Hunt, R. C. (2000). Rural versus Urban motor vehicle crash death rates: 20 years of FARS data. *Prehospital Emergency Care, 4*(1), 7–13.
4. Cenex. (2017). *What does more mileage really mean for vehicle life?*. Retrieved November 29, 2019, from https://www.cenex.com/about/cenex-information/cenexperts-blog-page/general-interest/what-does-more-mileage-really-mean-vehicle-life.

5. Chen, H. (2006). Applications of fuzzy logic in data mining process. In *Advanced Fuzzy Logic Technologies in Industrial Applications*, (pp. 249–260). London: Springer.
6. Clark, N. A., Demers, P. A., Karr, C. J., Koehoorn, M., Lencar, C., Tamburic, L., & Brauer, M. (2010). Effect of early life exposure to air pollution on development of childhood asthma. *Environmental Health Perspectives, 118*(2), 284–290.
7. DeJoy, D. M. (1989). The optimism bias and traffic accident risk perception. *Accident Analysis & Prevention, 21*(4), 333–340.
8. Department for Transport. (2019a). Vehicle licensing statistics: 2019 quarter 2 (April–June). Department for Transport. Retrieved November 25, 2019, from https://assets.publishing.ser vice.gov.uk/government/uploads/system/uploads/attachment_data/file/830795/vehicle-licens ing-statistics-april-to-june-2019.pdf.
9. Department for Transport. (2019b). Reported road casualties in Great Britain: 2018 annual report. Department for Transport, p. 1. Retrieved November 17, 2019, from https://assets.pub lishing.service.gov.uk/government/uploads/system/uploads/attachment_data/file/834585/rep orted-road-casualties-annual-report-2018.pdf.
10. Driving Standards Agency (2004). *The Highway Code*. The Stationery Office.
11. Evans, L. (1991). Traffic safety and the driver. Science Serving Society.
12. European Road Safety Observatory. (2018). *Annual Accident Report 2018*. European Road Safety Observatory. Retrieved December 9, 2019, from https://ec.europa.eu/transport/road_s afety/sites/roadsafety/files/pdf/statistics/dacota/asr2018.pdf.
13. Gainewe, M. A. G. A. D. I., & Masangu, N. D. E. N. G. E. Z. A. (2010). Factors leading to fatal crashes and fatalities on the South African roads: 2005–2009. SATC 2010.
14. Ghaemi, S., Khanmohammadi, S., & Tinati, M. (2010). Driver's behavior modeling using fuzzy logic. *Mathematical Problems in Engineering*.
15. Gov.uk. (2019). *Driving in adverse weather conditions (226–237)—The highway code—Guid- ance—GOV.UK*. Retrieved November 29, 2019, from https://www.gov.uk/guidance/the-hig hway-code/driving-in-adverse-weather-conditions-226-to-237
16. Hamam, A., & Georganas, N. D. (2008). A comparison of Mamdani and Sugeno fuzzy inference systems for evaluating the quality of experience of Hapto-Audio-Visual applications. In *2008 IEEE International Workshop on Haptic Audio Visual Environments and Games* (pp. 87–92). IEEE.
17. Hansen, B. K., & Riordan, D. (2000). Weather prediction using case-based reasoning and fuzzy set theory (Doctoral dissertation, DalTech).
18. Henning-Smith, C., & Kozhimannil, K. B. (2018). Rural-urban differences in risk factors for motor vehicle fatalities. *Health Equity, 2*(1), 260–263.
19. Holve, R., Protzel, P., & Naab, K. (1996). Generating fuzzy rules for the acceleration control of an adaptive cruise control system. In *Proceedings of North American Fuzzy Information Processing* (pp. 451–455). IEEE.
20. Hoyt, H. (1939). *The structure and growth of residential neighborhoods in American cities*. US Government Printing Office.
21. Ko, S. J., & Lee, J. J. (2007). Fuzzy logic based adaptive cruise control with guaranteed string stability. In *2007 International Conference on Control, Automation and Systems* (pp. 15–20). IEEE.
22. Krisciunas, K., & Carona, D. (2015). At what distance can the human eye detect a candle flame?. arXiv:1507.06270.
23. Kweon, Y. J., & Kockelman, K. M. (2003). Overall injury risk to different drivers: Combining exposure, frequency, and severity models. *Accident Analysis & Prevention, 35*(4), 441–450.
24. Lankarani, K. B., Heydari, S. T., Aghabeigi, M. R., Moafian, G., Hoseinzadeh, A., & Vossoughi, M. (2014). The impact of environmental factors on traffic accidents in Iran. *Journal of Injury and Violence Research, 6*(2), 64.
25. Mahowald, M. W. (2000). Eyes wide shut: The dangers of sleepy driving. *Minnesota medicine, 83*(8), 25–30.
26. Malik, M., & Majumder, S. (2013). An integrated computer vision based approach for driving assistance to enhance visibility in all weather conditions. In *International and National Conference on Machines and Mechanisms*.

27. Mamdani, E. H. (1974). Application of fuzzy algorithms for control of simple dynamic plant. In *Proceedings of the Institution of Electrical Engineers* (Vol. 121, No. 12, pp. 1585–1588). IET.
28. McConnell, R., Islam, T., Shankardass, K., Jerrett, M., Lurmann, F., Gilliland, F., et al. (2010). Childhood incident asthma and traffic-related air pollution at home and school. *Environmental Health Perspectives, 118*(7), 1021–1026.
29. Mende, J. (1982). Analysis of snow storm-related accidents in metropolitan Toronto. In *Proceedings of the 7th Annual Conference on Cost Effective Measures for Transport Improvements.* Chelsea Inn, Toronto, Canada. Retrieved May 30 to June 2, 1982.
30. Moore, G. E. (1965). Cramming more components onto integrated circuits.
31. Naranjo, J. E., González, C., García, R., & De Pedro, T. (2006). ACC+ Stop&go maneuvers with throttle and brake fuzzy control. *IEEE Transactions on Intelligent Transportation Systems, 7*(2), 213–225.
32. Naranjo, J. E., González, C., Reviejo, J., García, R., & De Pedro, T. (2003). Adaptive fuzzy control for inter-vehicle gap keeping. *IEEE Transactions on Intelligent Transportation Systems, 4*(3), 132–142.
33. Owsley, C., & McGwin, G., Jr. (2010). Vision and driving. *Vision Research, 50*(23), 2348–2361.
34. Park, R. E., Burgess, E. W., & McKenzie, R. D. (1967). *The city (1925)* (p. 1). Chicago: Chicago UP.
35. Parliament, E. (2010). Directive 2010/40/EU of the European Parliament and of the Council of 7 July 2010 on the framework for the deployment of intelligent transport systems in the field of road transport and for interfaces with other modes of transport text with EEA relevance. *The Official Journal of the European Union,* 1–13.
36. Perera, L. P., Carvalho, J. P., & Soares, C. G. (2013). Solutions to the failures and limitations of Mamdani fuzzy inference in ship navigation. *IEEE Transactions on Vehicular Technology, 63*(4), 1539–1554.
37. Rac.co.uk. (2017). Retrieved December 1, 2019, from https://www.rac.co.uk/drive/advice/learning-to-drive/stopping-distances/.
38. Road Safety Foundation. (2010). Protect & survive: star rating england's trunk road network for safety. Basingstoke: Road Safety Foundation. Retrieved November 29, 2019, from https://34.250.94.66/wp-content/uploads/2017/05/starrateenglandroadsweb1.pdf.
39. Rudny, D. F., & Sallmann, D. W. (1996). *Analysis of accidents involving alleged road surface defects (i.e., shoulder drop-offs, loose gravel, bumps and potholes)* (No. 960654). SAE Technical Paper.
40. Simões, J., Jaco, D., Gomes, R., Araujo, P., Fernandes, A., & Seabra, E. (2016). Evolution of the cruise control. In *International Conference on Regional Triple Helix Dynamics (HELIX2016).* Instituto Politécnico de Castelo Branco (IPCB).
41. Singh, S. (2015). Critical reasons for crashes investigated in the national motor vehicle crash causation survey. DOT HS 812 115. National Highway Traffic Safety Administration, US Department of Transportation.
42. Teetor, R. R. (1950). Speed control device for resisting operation of the accelerator. U.S. Patent 2,519,859.
43. Tesla.com. (2019). Autopilot. Retrieved November 25, 2019, from https://www.tesla.com/en_GB/autopilot.
44. Who.int. (2012). WHO | World Health Organization. Retrieved November 25, 2019, from https://www.who.int/ith/other_health_risks/injuries_violence/en/.
45. World Health Organization. (2018). *Global status report on road safety 2018.* World Health Organization.
46. Zadeh, L. A. (1965). Fuzzy Sets. *Information and Control, 8*(3), 338–353.
47. Zadeh, L. A. (1975). The concept of a linguistic variable and its application to approximate reasoning –I, I1 and 111, *Information sciences,* –01.8, pp.199–249, pp.301–357 and Vol. 9, pp. 43–80.

Automatic Camera Flash Using a Mamdani Type One Fuzzy Inference System

Sophie Hughes and Arjab Singh Khuman

Abstract Photography is an enjoyable hobby for many people, with many systems having been developed to make it easier for newcomers to begin learning how to take a quality photograph. Features such as automatic aperture and shutter speed allow the user to take a photo without any prior knowledge as to how these two should be manipulated in order to take a good photo. However, a feature that has not currently been explored is an automatic camera flash that will change intensity based on a number of factors, as current automatic flash systems will simply either activate a flash or not based on the perceived light levels of the image. This chapter will utilise a Mamdani type one fuzzy inference system in order to demonstrate how an automatic camera flash could potentially work, justifying each input used as well as discussing any limitations and possible improvements.

1 Introduction

Photography has become a very accessible hobby in recent decades, particularly due to the advent of digital cameras making it so that most of the work behind adjusting the settings of a camera (i.e. aperture and shutter speed) can handled done automatically. Many consumer camera will also be equipped with feature that allows the user to gain more control over these settings as they get to grips with the impact that each one has on a photo, namely, through settings such as aperture priority or shutter speed priority which allow the user to control the specified setting themselves, while the camera will automatically adjust the other in an appropriate way.

An aspect of photography that is heavily utilised among those more advanced in the discipline is lighting and its effect on a photograph. It is very common for consumer cameras to have a flash which will simply either fire or not, with no

S. Hughes
School of Computer Science and Informatics, De Montfort University, Leicester, UK

A. Singh Khuman (✉)
School of Computer Science and Informatics, Institute of Artificial Intelligence (IAI), Leicester, UK
e-mail: arjab.khuman@dmu.ac.uk

© Springer Nature Switzerland AG 2021
J. Carter et al. (eds.), *Fuzzy Logic*,
https://doi.org/10.1007/978-3-030-66474-9_13

221

variance in the intensity of that flash. This can often result in photographs with lighting that is too intense, resulting in an unintentionally harsh appearance to the resulting image. A user more experienced in photography will utilise flash and/or continuous lighting, adjusting each light source to be at an appropriate level in order to achieve the desired result. For many amateur photographers, this level of control may be unnecessary or overwhelming, which is where an automatic version of this dynamic lighting may prove to be useful.

The system that will be developed in this chapter will aim to bridge this gap, particularly for users who feel that a standard flash no longer produces photographs at a satisfactory quality, but who may also be intimidated by, or still learning how to use, a more professional manually adjustable flash. The results of using an automatically adjusted flash such as the one that shall be developed in this chapter will not be as optimal as those produced by a flash that has been manually adjusted by a knowledgeable user; however, it should prove to be sufficient for the average amateur user until they feel confident enough in their knowledge and abilities and can accurately and quickly adjust the light sources themselves.

2   Literature Review

Kaskowitz [3] define fuzzy logic as 'approximat[ing] human reasoning, by allowing ambiguity in membership sets. It is possible for something to belong to a particular set to a certain degree. In addition, it is likely that something will belong to several different sets to varying degrees at the same time'. This contrasts classical set theory, which uses Boolean logic and states that an object either has full membership to a given set, or none at all [4].

Many aspects of photography involve vagueness in some form, which makes fuzzy logic a good tool to solve many different problems across the discipline, such as solutions for auto-focus, auto-exposure and white balance. However, there has not yet been any work done in the area of developing a fuzzy logic-based camera flash. This review will cover other camera solutions and systems previously mentioned in order to investigate how their findings could contribute to the development of a fuzzy logic-based automatic camera flash.

2.1   Auto-Focus

Auto-focus, as the name suggests, is a system for automatically focussing the camera lens on the presumed subject. Kaskowitz [3] briefly describe how a fuzzy system could be used to determine which item in frame is the subject of the photograph and therefore should be focussed on. It would work by splitting the screen into three sections, left, middle and right, and determining which is closest to the camera, thus deciding that this is the subject of the photograph.

Avenue et al. [1] describe in more detail a fuzzy system for controlling auto-focus whereby the average grey level (ag) of the image and the rate of change of fuzzy index of the image are taken as inputs and iris control (ir) is taken as the output. The system takes the approach of 'if it is an edge, it should be sharp' and works by running the fuzzy calculations, adjusting the lens accordingly, and then repeating the process each time the maximum fuzzy index is attained until the output remains steady. It is noted that this implementation ends up being a form of feedback control.

Although designed for a video camera, Haruki and Kikuchi [2] also describe a fuzzy logic-based auto-focus. This system uses the detail and luminance of the objects in the shot in order to determine which is the subject of the photograph. The image is sectioned off into shaded areas which could potentially be the subject, and then each of these areas is divided into 16 subsections. Fuzzy rules are then ran which look at how detailed each subsection is, and the difference in luminance across the 16 subsections. If the subsections are detailed and the differences in luminance are high, then it is likely that the subject of the photo is within that section, and thus the camera should auto-focus to there.

All three auto-focus systems propose interesting ways to approach the problem, with each solution providing a different methodology in order to calculate where the camera should auto-focus. A system that is a hybrid of all three would likely produce the best results; however, with a system that complex, there could potentially be performance issues that would need to be solved in order for the camera to focus at an acceptable speed.

2.2 Auto-Exposure

Haruki and Kikuchi [2] also describe a solution for auto-exposure that uses fuzzy logic which works in a similar manner to their proposed solution for auto-focus. The image is divided into its major sections and then the luminance areas of each section are compared by a fuzzy inference system. If certain sections are close in luminance and others aren't, then specific other sections have their weighting increased. Other rules are also utilised such as if the maximum luminance is not low but the average is, then the darker areas of the image are brightened. It is noted that this system allows for a good amount of flexibility and adaptability to many situations.

A different solution is proposed by Shimizu et al. [5], where a new parameter called 'HIST' is used, which is defined as 'the ratio between the number of pixels whose brightness is higher than a threshold value and the total number of pixels in the whole TV picture'. The system measures three HISTs within three areas of the image; the mean value, half of the mean value, and twice of the mean value. These are then used in order to measure the contrast of each area of the subject and the background, to produce a value H_diff which is then used as an input to a fuzzy system alongside the mean value for HIST. The output for this fuzzy system is the compensation value for the brightness of that area of the image. In general, the rules used state that the system will only end up making compensations to areas which do not have a large H_diff value.

2.3 Conclusions

It is clear that fuzzy logic has been used in very novel and effective ways across photography and videography systems in order to improve the images produced. Many systems analysed have taken the approach of doing digital analysis of the image produced and then adjusting factors in order to improve that image. In the case of developing a fuzzy logic-based automatic camera flash, a different approach of analysing the current conditions and adjusting the flash intensity based on those conditions would be a more suitable approach as opposed to the feedback loop approach taken by a number of systems in existing literature. This would allow the system to maintain the 'point and shoot' appeal that automatic cameras have.

The papers in this review have had a common theme of having systems for multiple different aspects of taking a photograph that works alongside each other to produce a good quality photograph. Thus, a fuzzy logic-based camera flash should also be designed so that it is non-intrusive and would work in harmony with the other systems presented in this review.

3 System Design

3.1 Overview

Although a wide range of camera flash intensities can be used in order to have different effects on a photo, this system will be designed as a 'one size fits all' implementation of a flash, operating with similar presumptions to the automatic modes of consumer cameras which purely aim to get an acceptable photo. Similar to the parameters controlled by automatic mode, there is potential for much better results when controlling the flash intensity manually; however, the intensity provided by this system should be suitable enough to produce a suitably lit photo.

In order to calculate what the intensity of the flash should be, the system will take the following inputs:

- Distance from Subject (Metres).
- Ambient Light Level (Lux).
- Camera Aperture (F-Number).
- Camera Shutter Speed (Seconds).

These inputs will then be fed into the Mamdani type one fuzzy inference system (FIS) in order to produce the output of 'Flash Intensity'.

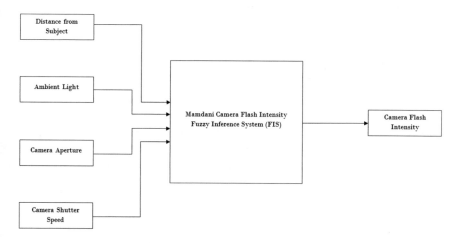

3.2 Input: Distance from Subject

Due to the fact that the light level hitting the subject decreases as the light travels from the flash, the distance between the flash and the subject needs to be taken into account. If the subject is very close to the flash, then less light will be lost by the time it reaches the subject. Conversely, if the subject is very far from the flash, a more intense light level will need to be produced in order for the subject to be sufficiently lit (Fig. 1).

If implemented into an actual product, this input could be measured by a sensor affixed onto the flash that would operate in a similar manner to laser measuring tapes.

The scale for this input begins at 0.1 m and ends at 5 m. A minimum value of 0.1 m was chosen as this is a very short distance and it would not be appropriate to use a flash at a distance smaller than this in consumer photography. A maximum value of 5 m was chosen as this is around the maximum distance where it would be appropriate to be using a camera mounted flash as lighting for the subject.

Fig. 1 The 'Distance from Subject' input, showing the membership functions

A Gaussian membership function has been chosen for this input as the perceived membership to each of the categories of distance (very close, close, far, very far) will change gradually as the input moves across the range.

3.3 Input: Ambient Light Level

The light that is already in the environment should be considered when deciding how intense the flash should be. In a brightly lit area, use of a very bright flash would not be necessary as opposed to in a very dark environment in which it would be impossible to capture an image without the use of a bright flash (Fig. 2).

If implemented into an actual product, this input could be measured by a light sensor affixed onto the flash.

Lux is chosen for the unit of measurement of this input as it is the SI unit of measurement for illuminance. For reference, around 500 lx is the average illuminance of an office, whereas 1000 lx is the illuminance of an overcast day. Thus, the scale for this input ranges from 0 lx (complete darkness) to 2000 lx (Illuminance required for doing fine detailed work).

A trapezoidal membership function has been chosen for this input as perception of brightness does not typically have sharp peaks. There are also large ranges of light levels that, in the context of photography, are considered to definitely fall within one of the three categories (dark, dim, bright). The property of a trapezoidal membership function having a plateau at its highest degree of membership allows for an accurate representation of this perception.

3.4 Input: Camera Aperture

The camera aperture is the measure of how wide the opening is that allows light to pass through to the image sensor. Adjusting the aperture will affect both how much light is allowed into the sensor (thus, how bright the image will be) and the depth of field of the image (i.e. how blurry the background is). The aperture is measured with an 'f-number' and a camera will typically have a set of 'f-stops' that the f-number can be set to. A small 'f-number' means that the lens opening is wide and a lot of

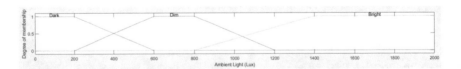

Fig. 2 The 'Ambient Light' input, showing the membership functions

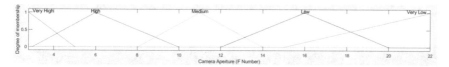

Fig. 3 The 'Camera Aperture' input, showing the membership functions

light will be let in, whereas a large 'f-number' means that the lens opening is narrow and not much light will be let in (Fig. 3).

A typical range of 'f-numbers' on a consumer camera is f/2.8–f/22, and thus this was chosen as the range for this input.

A triangular membership function has been chosen for this input as membership to a given category of aperture will move uniformly as the aperture changes, as opposed to a Gaussian membership function where the rate of membership to a given category will change as you move across the range.

3.5 Input: Camera Shutter Speed

The camera shutter speed is the measurement of the length of time that the image sensor is exposed to light. Thus, the longer the duration of the shutter speed, the more light will be allowed to hit the image sensor, resulting in a brighter image. A fast shutter speed will result in a sharper picture (as there will be less time for the camera or subject to move while the image sensor is exposed); however, it will also not let in much light. In order to get a sharp, bright photo, a fast shutter speed should be used in combination with a bright flash. Alternatively, if the shutter speed is slower (for example, if doing long exposure photography in order to achieve a deliberate blurred effect) then the camera flash does not need to be quite as intense (Fig. 4).

When using flash lighting, often the shutter speed is not as important as the other factors mentioned previously and can instead be used creatively as opposed to being used to achieve appropriate brightness in the photo. Because of this, the system has been designed so that the shutter speed has a lesser impact on the output in comparison to the other three inputs.

The shutter speed of a camera is measured in seconds (usually fractions of a second) with the standard range being 1/1000–1 s. Many cameras will be equipped

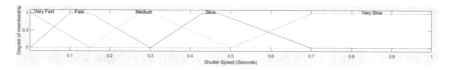

Fig. 4 The 'Camera Shutter Speed' input, showing the membership functions

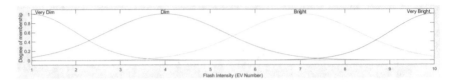

Fig. 5 The 'Flash Intensity' output, showing the membership functions

to have shutter speeds slower than this, often up to a minute long, however for the purposes of this system, up to 1 s should suffice.

A trapezoidal membership function has been chosen for this input as it allows for uniform changes in membership to certain categories, while still allowing for a level portion of a maximum degree of membership to a given category, notably in the 'Very Slow' category, where it is implied that every shutter speed past the maximum shown of 1 s would also be considered to be 'Very Slow'.

3.6 Output: Flash Intensity

The output for this system is the flash intensity. Unlike traditional light bulb flashes, modern electronic flashes can have their intensity adjusted, which forms the basis of this system. The level of intensity is commonly measured in 'EV Number' which serves as a consistent way for the photographer to know the flash output across different flash units. Each EV Number represents double the intensity of the number before it, for example, EV10.0 is defined as 6400 Ws, whereas EV9.0 is 3200 Ws (Fig. 5).

An EV Number range of 1–10 is common for consumer camera flashes, so that range has been chosen to be used at the output for the FIS.

A Gaussian membership function has been chosen as it would not be appropriate for the peaks of the membership functions to be as sharp as when using a triangular membership function, and the transitions between them should be relatively smooth.

4 The Rule Base

Initially, a list of 300 rules was drawn up, covering every combination of inputs for the system. This was tested with 250 randomly generated sets of inputs in order to confirm that a suitable spread of combinations of input values was accounted for. However, due to the amount of rules in this initial rule base, it would be very difficult to tune the rules so that they produce an accurate, expected output. It is, therefore, necessary to undertake a process of reducing the amount of rules in the system while ensuring that accuracy or correctness is not compromised.

Due to many of the rules being very similar, the first step taken to slim down the rule base was to analyse which sets of rules could be condensed into a single rule, with 0 taking the place of certain inputs. For example, the below rule would produce an output of 'Very Dim' if the Distance from Subject was 'Very Close', regardless of what the other inputs are.

$$exampleRule = [1\,0\,0\,0\,1\,1\,1];$$

An initial analysis identified 70 rules that could be made redundant due to their inputs being extremely similar and their outputs being the same, and so these were condensed into 13 rules for a total of 243 rules. Another 250 randomly generated input sets were then tested against these rules to confirm that they still covered a suitable spread of input combinations.

In this state, the rules treat all four inputs with the same importance. As mentioned previously, the shutter speed of the camera should not have as much of a bearing on the output as the other three inputs. Thus, the shutter speed was removed as an input to the existing rules and a set of rules were created to exclusively deal with the shutter speed input. These new rules were given a lower weighting of 0.7 while the rest were kept at 1. This weighting should allow these rules to have enough influence for the output to be sufficiently affected, but not as much as the other three inputs.

$$ssRule1 = [0\,0\,0\,1\,4\,0.7\,1];$$
$$ssRule2 = [0\,0\,0\,2\,3\,0.7\,1];$$
$$ssRule3 = [0\,0\,0\,3\,2\,0.7\,1];$$
$$ssRule4 = [0\,0\,0\,4\,2\,0.7\,1];$$
$$ssRule5 = [0\,0\,0\,5\,1\,0.7\,1];$$

Once the shutter speed had been removed as a variable from the main set of rules, the remaining rules were able to be condensed further into 60 rules. Aside from allowing the shutter speed to have a lesser impact on the results, the heavily reduced total of 65 rules also makes it much easier to tune said rules to produce a proper input. Again, another 250 randomly generated input sets were then tested against these rules to confirm that they still covered a suitable spread of input combinations.

5   Testing Defuzzification Methods

The last step of the fuzzy system is defuzzification, in which the resulting fuzzy set is translated into a crisp value. In the context of the camera flash system, this crisp value is the EV Number that the flash should be set to. There are a number of different methods of defuzzification, thus it is necessary to test the system using a variety of defuzzification methods in order to determine which is the most appropriate for the given system.

The testing strategy undertaken was to run tests that cover the extremes of each input. This method is efficient at highlighting any issues, as the extremities are more likely to be incorrect than middling values. A list of 40 test cases was drawn up and was fed into the FIS, changing the defuzzification method to the method to be tested each time. Below are a few of the tests to serve as an example.

Test number	Distance from subject (m)	Ambient light (lx)	Aperture (F-number)	Shutter speed (s)	Expected output	Actual output	Pass or fail?
1	0.1	2000	2.8	1	Very dim		
2	5	0	22	0.001	Very bright		
3	2.4	700	11	0.001	Bright		
4	2.4	700	11	1	Dim		
5	2.4	700	11	0.3	Dim		

5.1 Defuzzification Method: Centroid

Centroid defuzzification returns the centre of the area underneath the curve of the fuzzy set, so that if the shape were to be cut out, it would balance exactly at that point.

5.2 Testing Centroid Defuzzification

The tests for this defuzzification method had a pass rate of 100% and the outputs produced are reasonably varied, suggesting that the outputs produced are appropriately bespoke for each of the given inputs.

5.3 Defuzzification Method: Bisector

Bisector defuzzification divides the area under the curve of the fuzzy set into two segments of equal area. It is quite similar to centroid, with the line produced sometimes being the same.

5.4 Testing Bisector Defuzzification

The tests for this defuzzification method had a pass rate of 95% and the outputs produced were varied in a similar manner to centroid.

5.5   Defuzzification Methods: LOM, MOM and SOM

LOM, MOM and SOM stand for Largest, Middle and Smallest of Maximum, respectively. If, for example, the maximum value of the combined membership function was a plateau, each of these three values would be distinct. However, if there was a single peak, all three would be identical.

5.6   Testing LOM Defuzzification

The tests for this defuzzification method had a pass rate of 72% and the outputs produced were not particularly varied, especially in comparison to those produced by centroid and bisector.

5.7   Testing MOM Defuzzification

The tests for this defuzzification method had a pass rate of 70% and the outputs produced were also not particularly varied however they were slightly more so than LOM defuzzification.

5.8   Testing SOM Defuzzification

The tests for this defuzzification method also had a pass rate of 70% and the outputs produced were also not particularly varied, with the variety being similar to the outputs produced by LOM defuzzification.

5.9   Conclusions from Testing Defuzzification Methods

It is clear that LOM, MOM and SOM are not suitable defuzzification methods for this system. The failure rate for the given tests was significantly higher than that of centroid and bisector, and the outputs themselves were much too repetitive to be acceptable. It is important to note that due to the nature of the majority of the test cases dealing with the combination of extreme values with middling values, the outputs produced across this range of tests would be expected to be somewhat similar due to the extremes being balanced out at a consistent rate by the middling values they were tested alongside. However, the outputs produced by LOM, MOM and SOM were very rigid and would not produce the dynamic, nuanced lighting system that is desired.

Centroid and bisector had more promising results, with the pass rate for those two being 100% and 95%, respectively. Compared to the low 70% pass rates that LOM, MOM and SOM produced, this instantly makes these two defuzzification methods much more suitable candidates for the defuzzification method to be used. When looking at the outputs produced by these two methods, it is clear which is the most suitable. Bisector produced a decent variety of outputs, and although not as extreme as what was observed with LOM, MOM and SOM, there is clear repetition in the outputs which, as mentioned previously, would not contribute well to producing a dynamic and fit for purpose system. On the other hand, centroid had very few repeated outputs, and groups of tests that often had repeated values in other methods ended up having a reasonable variety when centroid was used. This variety, on top of being the only method to have a pass rate of 100%, shows that centroid is the most suitable method of defuzzification for this system.

6 Critical Reflection

The next logical step for this system would be to test using a real flash, as explained in the system overview, given the correct hardware it is very achievable for a flash to be produced that utilises this system. The membership functions, rule base and defuzzification methods could then be fine tuned based on that practical testing. The system in its current state has been designed based on photography fundamentals and principles and although it proves as a good baseline, it is likely that there exist slight imperfections that could be improved upon if it were to be implemented and tested in a real flash. An iterative approach to testing could be taken in which the flash is tested under various scenarios (i.e. with various inputs) and then the resulting photographs are reviewed in order to determine if the light level was appropriate. Adjustments to the system could then be made and then the flash tested once again. This process would be repeated until it is confirmed that the resulting photographs consistently have an appropriate light level.

In the current state, the only input that has a different weighting from the rest is the shutter speed. There could potentially be a justification to follow a similar manner as described above in order to isolate the rest of the inputs so that their weightings can also be adjusted separately. A potential next step could be to isolate the 'distance from subject' input and reduce the weighting slightly, but not as low as the weighting of the shutter speed, and then evaluating whether or not it would be beneficial to follow this process for the final two inputs. By doing this for all the inputs, and testing the system with a real flash as described above, the system could potentially be very precisely optimised to the point where it could be implemented into a product for consumer use.

The system also doesn't currently make use of any 'or' rules (for example, if the distance from the subject was high OR the light level was low, then produce a given output), so it is likely that through further analysis and testing, cases could potentially be identified in which it would be beneficial to have rules which utilise this operator which could further serve to reduce the number of rules in the system.

7 Conclusion

The fuzzy inference system produced in this chapter serves as a baseline to a system that could potentially be developed further to the point of being able to be implemented into a viable consumer product. As mentioned in the critical reflection, there are a number of tweaks that could be made to the system as it currently stands. However, the largest leaps in improving this system would come from implementing this system in a practical setting in order to observe how it would perform and adjust the system accordingly. Any further adjustments to the system in its current state could potentially be proven to be unnecessary by testing in a practical environment, and so it would be sensible to withhold any further changes until some practical testing has been carried out in order to evaluate how the flash currently performs.

As highlighted due to the findings from the literature review, optimally this flash should be able to work alongside existing fuzzy logic-based systems commonly used in cameras. The inputs and outputs of the system that has been developed are not influenced by or would influence any inputs or outputs for the systems analysed in the literature review; therefore, the system achieves its goal of being interoperable with existing camera systems.

References

1. Avenue, T. C., Kong, H., Polytechnic, H. K., & So, A. T. P. (1993). Implementation of fuzzy logic based automatic camera focuser. *Search*, 292–295.
2. Haruki, T., & Kikuchi, K. (1992). Video camera system using fuzzy logic. *IEEE Transactions on Consumer Electronics, 38*(3), 624–634. ISSN 00983063. https://doi.org/10.1109/30.156746.
3. Kaskowitz, M. (1993). Consumer applications of fuzzy logic in video cameras. In: *Digest of Technical Papers—IEEE International Conference on Consumer Electronics (number Figure 2)* (pp. 238–239). ISBN 0780308433.
4. McKinsey, J. C. C., & Boole, G. (1950). The mathematical analysis of logic. *The American Mathematical Monthly, 57*(5), 351. ISSN 00029890. https://doi.org/10.2307/2306226.
5. Shimizu, S., Kondo, T., Kohashi, T., Tsuruta, M., & Komuro, T. (1992). A new algorithm for exposure control based on fuzzy logic for video cameras. *IEEE Transactions on Consumer Electronics, 38*(3), 617–623. ISSN 00983063. https://doi.org/10.1109/30.156745.

The Application of Fuzzy Logic in Determining Outcomes of eSports Events

Spencer Deane and Arjab Singh Khuman ⓘ

Abstract As eSports skyrocket in popularity, the saturation of top talent intensifies. Hundreds of millions of dollars in prize money are distributed amongst this talent, resulting in fierce competition. To get ahead, players go to extreme measures to gain marginal performance increases. Besides intense training and performance enhancing drugs, athletes seek intelligent analytical tools which can provide useful insights into a player's strengths and weaknesses. This report showcases a fuzzy system which uses real-world data and determines a player's percentage chance of winning a duel in the online first-person shooter video game Counter-Strike Global Offensive, one of the leading eSports.

Keywords Fuzzy logic · Esports analysis · Esports performance · Mamdani inference system · Counter strike global offensive

1 Introduction

This paper demonstrates a Fuzzy Logic system which determines the success chance of a Counter-Strike Global Offensive eSports player in a one versus one duel. The general definition of Counter-Strike Global Offensive and that of a duel are explained in detail later. Taking in a variety of data sets, an accurate system should present a percentage which maps to the player's genuine likelihood of succeeding against an opponent. Using this system players can enter their own data in an effort to discover their strengths and weaknesses and influence their future behaviour to increase their success. This system also serves as a demonstration of simplified chaos and proves that the potential for commercial analytical tools in eSports is truly in its infancy.

S. Deane
School of Computer Science and Informatics, De Montfort University, Leicester, UK
e-mail: P17165184@alumni365.dmu.ac.uk

A. S. Khuman (✉)
School of Computer Science and Informatics, Institute of Artificial Intelligence (IAI), Leicester, UK
e-mail: arjab.khuman@dmu.ac.uk

© Springer Nature Switzerland AG 2021
J. Carter et al. (eds.), *Fuzzy Logic*,
https://doi.org/10.1007/978-3-030-66474-9_14

Utilizing Boolean logic makes it difficult to create systems which rely on given points across a spectrum. Fuzzy logic fills this gap and provides the framework to allow for "degrees of truth" as opposed to "binary truth". This is necessary for eSports analytics, as the input variables are often taken as a range from one number to another. Fuzzy logic also requires us to build standardized systems of measurement which improve the reliability and universal acknowledgement of certain statistics. For example, measuring a player's aim and outputting a single score to represent how good their aim is may at first seem challenging to calculate, however, by creating standardized systems and equations, we can in fact calculate this score and offer future academics the ease of access to these methods, resulting in greater productivity over time.

The system was created using the Fuzzy Inference System within MATLAB. MATLAB offers rapid prototyping of finished products and provides a great amount of accuracy and control over the coding process. The syntax and structure of the code is easy to pick up versus more complicated languages such as C++, giving even beginners the opportunity to alter and extend the system. No external programs or API's are required to execute the code and build the system making it a neat and portable solution for generating Mathematical models at pace.

It is important to explain the game of Counter-Strike and to provide a general definition of a duel. Within competitive Counter-Strike there are 2 teams of 5 players each. In a single game, the winner is the first team to achieve 16 round wins. For the first 15 rounds, one team plays as Terrorists, the other as Counter-Terrorists. After 15 rounds the sides are switched. It is the Terrorist's job to either kill the entire enemy team or to plant and explode a bomb at one of the designated bomb sites. After a certain time, the bomb that is planted will explode, so it is the Terrorist's job to guard the bomb after it has been planted. As you might imagine, stopping the Terrorists from entering the bomb sites is a critical role for the Counter-Terrorists, and numerous strategies make the game extremely complex. At the start of each round, players can buy a gun, grenades and armour. These all have a huge impact on the outcome of the game, and economic strategies often come into play to ensure maximum impact.

A duel can be defined as one versus one interaction between two players on opposing teams. Given the variance in guns, grenades and armour, it is likely that one player will be out matched, and especially likely since often a player is taken off guard or is simply in a worse position. By shooting the enemy, the player can reduce their health, and when it reaches 0 the enemy dies.

There are many scenarios within Counter-Strike that are too complicated for current systems. If the player is currently a Terrorist and is pushing into a bomb site, they may be at a strategic advantage, due to the way the map has been designed. It is difficult to quantify this as a strict value. If a Counter-Terrorist decides to use a certain position, the enemy might suspect the player is in this position, and therefore, have a huge advantage. It could be argued that this is an unrealistic expectation for an analytical tool to make, since only the player can know what they are thinking.

Given the numerous potential factors that might impact the player's success chance, it is easy to see why such a complicated analytical system is needed, and why a satisfactory one has not been created so far.

2 What is eSports?

"eSports" is the term given to competitive online video gaming. The only requirements to be an eSport are players and an audience. The most popular eSports range in genre, for example, League of Legends and Dota 2 involve teams battling as avatars in arenas, whereas games such as Counter-Strike Global Offensive and Fortnite involve the player shooting the enemies with a gun. Most successful eSports contain a variety of strategies, quick thinking and teamwork. Awarding over $211 million in prize money in 2019 alone [1], eSports competitions offer millions of people across the world entertainment paramount to sporting events. As with most sporting events, eSports gambling is offered by numerous betting sites, and scandals involving players throwing matches for personal gain paint a picture of the popularity and financial potential in the eSports industry.

Total eSports viewership is expected to grow at a 9% annual compound growth rate between 2019 and 2023, up from 454 million in 2019 to 646 million in 2023 [2]. eSports investments increased 837% from 2017 to 2018, from $490 million to $4.5 billion [2]. With the money generated from ticket sales, merchandizing, sponsorships and many other streams of revenue, it is painstakingly clear that the magnitude of intensity players are willing to put into the game has never been greater. If the ubiquity of analytical tools in traditional sports holds up in eSports, the demand for them will vastly outpace the developer's ability to create them.

In 2019, the Sports analytics market was valued at $788 million and is expected to reach $3.07 billion by 2024 [3]. Observing the astronomical figures, assessing players and teams with a numbers-based approach appears to be an extremely valuable asset that owners of teams are extremely interested in, and given the high degree of similarity between traditional sports and eSports, it isn't unimaginable that eSports will match the pace. The application of each tool is very different, for example, certain systems may crunch the numbers of wins and losses and predict who will win, a more sophisticated system may suggest improvements that could be made, and an even more sophisticated system might expose flaws in teams that wouldn't be humanly possible due to the limited nature of our thinking.

Real-time competition often comes down to fractions of seconds, and training for an extra hour or discussing tactics for an extra ten minutes can easily be the deciding factor between winning and losing. Given the high stakes that have been discussed, it is in the player's best interest to optimize and refine their ability to compete.

It is also worth mentioning the cascading effect that the projected growth of the eSports industry will have. As prize money increases players will train harder and longer, which in turn, improves the viewing experience for fans, resulting in the ability for higher ticket sales and increased merchandising. Given that in 2018, only 1.43 billion people were aware of eSports [4], we can make assumptions about how this cycle will continue for years to come until it hits its saturation level.

3 Literature Review

The extensive application of fuzzy logic provides an abundance of knowledge and research that can be utilized to determine the validity and usefulness of a system like the one this paper demonstrates. Given that useful inference systems have been demonstrated in areas such as strength training [5], Artificial Intelligence [6] and political election results prediction [7], it's easy to understand that vagueness and uncertainty can demystify human interpretation into understandable results.

Bridge, along with chess, are "the only mind games officially recognized by the International Olympic Committee" [8], and whilst the performance analysis of Bridge players may seem like a mismanagement of fuzzy systems, it provides us with an insight into how commandingly simple data input needs to be for us to understand the landscape of potential growth opportunities in a players success. In "Assessing the Players' Performance in the Game of Bridge: A Fuzzy Logic Approach" Bridge data was used to determine factors such as which teams were better, which players were better, which could then be used to extrapolate subsets, for example, whether men were better than women. This is especially important to us as using the data we obtain to reach conclusions that can help the Counter-Strike player increase their performance. In the paper, real Bridge players were used to demonstrate the capability of the system, however, the sample size was only small. This is largely a non-issue as the purpose of the paper was only to demonstrate the system's potential. The paper also clearly demonstrates the method using graphical elements, as well as descriptions of the defuzzification process, which will be discussed later.

A similar analysis was performed on Cricket players [9]. The system used in this paper perfectly demonstrates the application of fuzzy logic in performance analysis. In the paper, inputs are taken such as "Runs Scored", "Balls Faced" and "Strike Rate", which are used to evaluate "the performance of a cricket player in batting and the impact of his performance on the ICC ranking" [9]. The paper makes use of several graphics to illustrate their ideas and methodology. Each aspect of the system is clearly laid out, including the rules used to determine how the inputs affected the outcome. A potential problem identified is that the system has 8 inputs and 96 rules. This means that every combination of the rule cannot possibly have its own outcome. For example, in the system demonstrated in this paper with 3 inputs, there are 27 rules. This also takes into account that for each input there are three membership functions, for example, "low", "medium" and "high". It is not explained in the paper how this problem was solved.

An understanding of team dynamics and player combinations is also relevant to analyzing a player's performance. "The success or failure of any team lies in the skills and abilities of the players that comprise the team" [10]. The paper demonstrates its method succinctly and uses graphical elements to aid in understanding. It is not explained as to the reason behind selecting certain membership functions, which can cause confusion for academics newer to Fuzzy Logic. It is possible that a framework like the one this paper provides could be used in the future as an addition to the system designed in this paper, which could be used to predict outcomes at levels never thought possible. Using the system in this paper, not only predicting a player's performance is valuable, but how valuable they are to the team is, of course, a relevant detail. Managers and team owners alike can obtain better readings of who they should be scouting or kicking to achieve greater results.

Another potential improvement to this paper's fuzzy system would be to incorporate holistic factors into its decision-making. Analyzing "Investigating the Human Factors in eSports performance" [11], we can see that several performance benchmarks such as dedication, practice, concentration, critical thinking and physical ability were compared between eSports and sports athletes. Certain aspects of the paper failed to hold up to the scientific method. For example, when interviewing coaches on the key human factors affecting performance, it is stated that there were no set questions asked to the coaches, but rather the same sort of questions were asked. While this may provide a sufficient result it is prone to error and given very few coaches were interviewed it seems likely that any given biases will likely have impacted the final result. The paper successfully explained their testing method and gave surveys to a large sample of people to achieve their results. The paper makes a meaningful conclusion and illuminates the variety of data required to totally and accurately advise a player on their performance gaps.

Real-time analytics holds an important part in eSports prediction, especially regarding the gambling scene. As the prevalence of analytical software utilizing fuzzy logic, artificial intelligence and machine learning grows, players will be forced to look at their performance through the eyes of a data scientist. It was discovered that prior to a match of Dota 2 (a popular eSport) the "Real-time eSports Match Result Prediction" [12] system had a 71.49% chance of predicting the winner. When using real-time features, this chance increased to 93.73%, indicating not only that real-time analysis is important, but that the events leading up to the game had a much greater impact than the single, isolated performance. The paper demonstrates their testing method well and provides a meaningful conclusion. It builds on previous work by providing additional inputs. This gives motivation to scholars and technically oriented people who believe everything is predictable providing you have enough mathematical data.

Using behaviour metrics such as "character class, weapon preference and play style preference" a framework was created to algorithmically determine which players would perform best when matched with other players [12]. Undeniably, it is almost always the case that the best players working together will outperform all other variants of player, however, the model was only demonstrated on the game "Destiny 2", a lesser known eSport. If the model were applied to games such as Counter-Strike, a game known for its heavy involvement in tactics and strategy, we may see unlikely pairs of players mapped together, who then go on to produce surprising and incredible results. For Counter-Strike, a new system would have to be designed, which could use the system laid out in this paper in conjunction with the system exhibited in "A Team Based Player Versus Player Recommender Systems Framework For Player Improvement" [12]. The paper does well to present information in a readable manner, providing the clustering methods where necessary and even describing the entire process of data manipulation. The system excellently demonstrates that the system created could be applied to a similar game, given enough adaptation. This is useful to us, as an adaptable system is a criterion for success when developing systems such as this.

"Esports Analytics Through Encounter Detection" [13] proves that "outcomes can be predicted based on the initial conditions, and the outcome of encounters be used to predict match outcomes" [10]. In this study, the eSport "Dota 2" was used. The win probability of a team was analyzed through many statistical lenses, for example, the amount of resources a team has, the encounter results and the kill differences. This was inputted to a combined model allowing us to see where the team needed to pick up the slack. The magnitude of importance that the use of this type of accumulative model is used is paramount. It is of very little importance to predict a player's win probability if you are unable to explain what went wrong. Whilst the system is built for MOBA's (Multi-player Online Battle Arena), the fundamental concepts demonstrate potential. An algorithm is explained which determines what can be defined as an encounter, and this is accompanied by an explanation of the game. This will guide our thinking on how best to put forward the ideas in this paper without confusing readers. As mentioned in the conclusion, the paper mentions that "Another approach to improve the outcome of encounters and matches is to increase the granularity of units to specific types of units. However, this approach requires a strongly increased database of matches". This highlights an important point about the current nature of analytics. The ability of individuals or teams of developers to obtain these large data sets is unlikely to happen for niche topics such as this one for several years.

In "An Open-Sourced Optical Tracking and Advanced eSports Analytics Platform for League of Legends" [14], it was shown that "advanced stats both better correlate with and explain team outcomes" [14]. It was also expressed that the results "suggest specific new ways forward for traditional sports" [14], and "provide a method for generating brand-new insight" [14]. In this paper, a win percentage was applied to a team (in the game "League of Legends") given an unyielding amount of data, ranging from minutes played, attack damage, movement speed, attack speed, critical damage, attack range, damage mitigated or shielded, stealth status and much more. This was then used to calculate a team win percentage. This is an impressive model and potentially the most accurate model eSports currently holds to display the strengths and weaknesses of players, and their contribution to their teams. Aside from being mathematically flawless, this paper visualizes its results using a variety of graphical elements. This easy to understand visual representation sets the standard for software developed using this model, since a casual observer must be able to understand the results with zero understanding of complex mathematics. Abstracting or removing certain data is necessary when building an accurate model—not only due to time and resource constraints, but very often certain data simply has no effect and shouldn't be muddied with the results. "Both casual observers and professional commentators often use the net gold difference between teams as an indicator of win probability. However, this is an inherently flawed approach, especially as the game progresses, as the team with less gold can often win a team fight, ace the enemies, and march down to destroy the nexus. Furthermore, total enemy gold is not revealed to players in game and must be estimated" [14]. Using open source libraries, machine learning models were used to scrape the screen to determine outcomes from the "mini-map" (a map detailing the movement of the players within the game). Another machine learning approach was used to interact directly with the game client to retrieve data through that pathway. Another method of machine learning was mentioned whereby the author would not have to interact with the game client but instead interact with the network directly. All of these methods offer ideas to other academics attempting to utilize machine learning within eSports. Using visual cues such as the mini-map, data can be obtained that typical players would build up over a life-time of playing. An example of how this could be used within Counter-Strike might be that when the enemy team is seen at location X, it is Y% likely that they will push onto the B bomb site, instead of onto the A bomb site. Using machine learning, insights can be gained which improve team performance by improving an individual's ability to respond to change. The methods used in this paper will aid in the future eSports analytics tools and will inform the thinking outlined in this paper as to the potential for eSports analytics tools.

It's clear that fuzzy logic has applications in the field of eSports. It has demonstrated that predicting the outcome of events is possible by developing a fuzzy model and plugging in real-world data. A Mamdani style fuzzy inference system is a sensible approach to determining the outcome of a duel in Counter-Strike and has the potential to produce accurate and useful results with real-world application, as shown in several of the reviewed papers.

4 System Design

The core goals of the fuzzy system are to process data relevant to the duel and produce the chance that the individual will win the duel as a percentage. Whilst the scope of this system is limited to the fundamental building blocks of a player's skill, the framework demonstrated can be adapted in many different ways. The system is laid out clearly enough so as to allow for the reproduction and adaptation of the system (Fig. 1).

Given the limited scope of the system, it is necessary to abstract certain data sets. The opposing player's skill, for example, is abstracted. This is mainly due to time constraints, however, over a million interactions the variety of opponents the player will have average out; similar to poker, an analysis of a player's performance versus a single opponent isn't telling, the performance of that player over a long time period paints a better picture of their skill level. Other aspects of the duel such as how much the player slept the night before are not included due to lack of information, however,

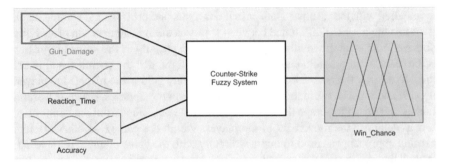

Fig. 1 The Mamdani fuzzy system

this information would be valuable in a more personalized system. The aim of this system is not to offer a commercial analytical tool but to build the foundations and expose the possibility of an analytical approach to eSports performance.

The first input is gun damage. Within Counter-Strike Global Offensive there are many guns a player can utilize. Each has its own statistics, for example, firing rate, bullet damage, accuracy range, etc. To avoid complications all aspects of the gun have been simplified to gun damage. Calculations between the firing rate and base bullet damage have been made to ensure a fair standard across all guns. This allows for guns such as shotguns to be used in our system; typically, a shotgun will only fire once and then a delay will stop the player from shooting. This again exposes the simplicity of the system as the play style (the way in which the player moves and acts using the gun) of the player is heavily impacted by the type of gun they use. If they are using a shotgun they might fire, take cover, peak out and fire again, whereas with a rifle they may just stand and shoot. So for the sake of demonstrating analytical potential these aspects have been abstracted in an appropriate manner.

The second input is reaction time. The formula to calculate this simply measures the point at which the enemy player can be seen to the time the player visibly reacts to the visual stimuli. Requiring athletes to perform a reaction time test proves futile for many reasons. Within Counter-Strike Global Offensive, visual stimuli are mixed with many other factors; a player with impeccable reaction time may falter in-game due to their inability to manage multiple data sources, such as the map of the level, or the stress of the game. This is why it is necessary to analyze a recording of the player's in-game performance and use a timer to calculate the reaction time. Potential improvements such as software-based reaction time trackers are possible, this could also be incorporated with eye tracking software which supplements the data.

The third input is accuracy. The standardized model for calculating accuracy has been built as a suggestion to readers for future analytical tools. It can be described simply by saying "how far from the desired target did the player's bullet land". There are many aspects within this that could be taken apart and utilized to give a more accurate reading. The specific recoil (how much the gun kicks up when it's fired) isn't considered, merely the player's ability to control the recoil and produce an accurate shot is. Showcasing this tool also deprives the potential of using the player's desired

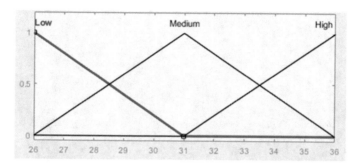

Fig. 2 The gun damage input variable

target as an input; in Counter-Strike Global Offensive shooting an opponent's head will do the most damage, followed by the chest. If the opponent is low health, then targeting the chest may be optimal as it is a bigger target. This ambiguity has been discounted due to the fact that in a duel, the players do not know their opponent's health. As well as this, the vast majority of shots are targeted towards the head, so this abstraction will only reduce the accuracy of our results slightly. Again, the aim of this system isn't to provide highly accurate analysis, but to provide groundwork for future tools, and to simply say that altering given inputs will have a clear effect on the output.

Each input uses "Trimf" membership functions. This means that three points are used to establish shape. Trimf was used due to the requirement of constant change in output. Other membership functions, such as "Gauss" are unable to meet this requirement due to the variable rate of change due to the changing gradient of its curves.

For the gun damage input, the minimum damage is 26 and the highest is 36. This is taken from the statistics of the official guns in the game. Three membership functions were given: low, medium and high. Additional membership functions were tested however they failed to add accuracy to the final result so were removed (Fig. 2).

Official data was used to interpret the minimum and maximum reaction times. The average human reaction time is about 0.25 s. Further, statistical analysis proves that this maps well to Counter-Strike Global Offensive, and system adaptations should use this value as the average. The minimum reaction time is set to 0; this a safety feature. Whilst it is evident no human can react in 0 s, setting the minimum reaction adds an element of risk to the system, as an unprecedented reaction time could break the system. The upper bound of the data has been designated as 0.5. After several rounds of testing symmetry within the data was necessary for providing an accurate output.

The system needed to account for very slow reaction times, however, it was important that after a certain point, the impact on the player's chance of winning didn't change (Fig. 3).

A range from 0 to 10 was given to the accuracy input, since this is a self-defined metric. The definitions for 0 accuracy and 10 accuracy are as follows: after gathering

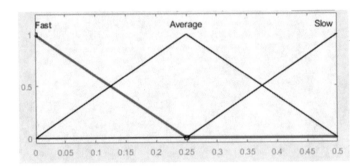

Fig. 3 The reaction time input variable

all the test data the furthest distance from the centre of the players head that a bullet landed acted as the lowest possible accuracy, and the closest a bullet landed to the centre of the player's head acted as the highest. An alternative approach could say that the centre of the player's head proves the best accuracy, and then a given range could be picked to say that anything outside that range is 0, and of course, anything within range would be mapped to a 0–10 accuracy range. Similarly, to the other inputs, low, medium and high membership functions were used (Fig. 4).

To obtain the testing data several recordings of professional tournament Counter-Strike games were scanned. To ensure fairness a variety of skill levels were picked, and in total 30 interactions were measured. The only criteria for a recording to be eligible is that the players are evenly matched. If they have different weapons then the result will be inaccurate. More advanced systems will consider the opponent. If one player doesn't know about the other player (for example, he is snuck up on from behind) then this discounts the recording as well. A variety of guns were picked. This may at first propose errors in the scientific method, as within Counter-Strike most players use the AK-47 or the M4, however for the purpose of this demonstration it is important to prove that the system works across the spectrum of guns, and is, therefore, less concerned with the realism of gameplay.

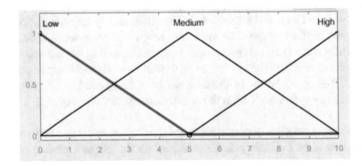

Fig. 4 The accuracy input variable

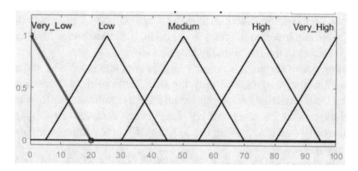

Fig. 5 The win chance output variable

Rigorous testing proved that five membership functions in the output yielded better results than three, as it allowed for enough variability in the outcome of individual rules. With only three it is impossible to distinguish between rules that should have different impacts on the final outcome. The trimf membership function was used because constant change in the output is paramount (Fig. 5).

Within MATLAB and Fuzzy Logic "rules" determine which inputs equate to which outcome. For example, a low gun damage, a slow reaction time, and a low accuracy will result in a low chance of winning. However, altering only the gun damage to be medium instead of low will now result in a higher win chance. As single alterations require impacts on the win chance our system must have a rule for every possibility. Every different scenario must have a separate rule which determines the outcome.

A flaw in this system and a problem for academics to solve is that even setting the rules requires a degree of subjectivity. For example, should moving from low gun damage to medium gun damage have the same impact on the win chance as a slow reaction time moving to a medium reaction time? The advancement of machine learning and artificial intelligence will prove useful in determining the weight and impact each variable can have within a system. Without utilizing this technology abstractions have been made which will produce a certain level of inaccuracy. In general, having low gun damage is less impactful than having a slow reaction time, and having a slow reaction time is less impactful than having low accuracy.

When testing the system a C++ script was used to determine the ranking of each rule. The algorithm which can achieve this is fairly rudimentary and so won't be fully discussed here, however, the underlying algorithm assigns a value to each input and then calculates the overall score of that sequence of rules. By doing this to every sequence of rules we can sort them to achieve a perfect ranking. This method has an obvious subjectivity when assigning values to each of the inputs, however, using the script is the best way to minimize subjectivity or error.

Within Fuzzy Logic, a defuzzification method is used to map fuzzy inputs to a crisp output. Examples are Centroid, Bisector, MOM, etc. Centroid provides a value that is central to the area of the output shape, and the output shape is determined by the strength of the value relative to the membership function. If the value is halfway up the slope of a membership function, then the shape will be the membership function cut in half. One can imagine the final shape being the tallest shape of a combination of the shapes filled in the membership functions.

The testing phase for the system was largely trial and error. The same data was used throughout the process, and only the rules and the inputs were altered. The criteria for success was that each input would have a significant impact on the output and would align with the rules entirely. After every iteration of the system, every variation possible was tested using the MATLAB toolbox to see which rules were not firing as expected.

During the testing phase, each defuzzification method was tested. In Fig. 6, the test data is categorized as belonging to one of its membership functions. It is apparent that this method will reduce the accuracy of the results, as a value on the high end of a function will be categorized the same as a value on the low end of a function,

Case	Gun Damage	Reaction Time	Accuracy	Win Chance Expected	Centroid	Bisector	MOM	LOM	SOM
1	MEDIUM	AVERAGE	MEDIUM	MEDIUM	MEDIUM	MEDIUM	MEDIUM	MEDIUM	MEDIUM
2	HIGH	AVERAGE	HIGH	VERY HIGH	MEDIUM	HIGH	VERY HIGH	VERY HIGH	VERY HIGH
3	LOW	FAST	HIGH	MEDIUM	MEDIUM	MEDIUM	MEDIUM	MEDIUM	MEDIUM
4	HIGH	AVERAGE	HIGH	VERY HIGH	HIGH	HIGH	VERY HIGH	VERY HIGH	VERY HIGH
5	MEDIUM	AVERAGE	MEDIUM	MEDIUM	HIGH	MEDIUM	MEDIUM	MEDIUM	MEDIUM
6	HIGH	AVERAGE	MEDIUM	HIGH	MEDIUM	HIGH	HIGH	HIGH	HIGH
7	MEDIUM	SLOW	MEDIUM	LOW	MEDIUM	MEDIUM	LOW	LOW	LOW
8	LOW	AVERAGE	HIGH	MEDIUM	MEDIUM	MEDIUM	LOW	MEDIUM	LOW
9	HIGH	SLOW	MEDIUM	MEDIUM	MEDIUM	MEDIUM	MEDIUM	MEDIUM	MEDIUM
10	MEDIUM	FAST	HIGH	VERY HIGH	MEDIUM	HIGH	HIGH	HIGH	HIGH
11	MEDIUM	AVERAGE	HIGH	HIGH	MEDIUM	HIGH	HIGH	HIGH	HIGH
12	MEDIUM	FAST	HIGH	VERY HIGH	HIGH	HIGH	VERY HIGH	VERY HIGH	VERY HIGH
13	HIGH	AVERAGE	HIGH	VERY HIGH	VERY HIGH	VERY HIGH	VERY HIGH	VERY HIGH	VERY HIGH
14	HIGH	AVERAGE	HIGH	VERY HIGH	VERY HIGH	VERY HIGH	VERY HIGH	VERY HIGH	VERY HIGH
15	MEDIUM	SLOW	MEDIUM	LOW	LOW	LOW	LOW	LOW	LOW
16	MEDIUM	FAST	MEDIUM	HIGH	HIGH	HIGH	HIGH	HIGH	HIGH
17	LOW	SLOW	LOW	VERY LOW	VERY LOW	VERY LOW	VERY LOW	VERY LOW	VERY LOW
18	MEDIUM	AVERAGE	MEDIUM	MEDIUM	HIGH	HIGH	MEDIUM	MEDIUM	MEDIUM
19	LOW	AVERAGE	LOW	VERY LOW	LOW	LOW	VERY LOW	VERY LOW	VERY LOW
20	HIGH	AVERAGE	MEDIUM	HIGH	HIGH	HIGH	HIGH	HIGH	HIGH
21	LOW	AVERAGE	MEDIUM	LOW	MEDIUM	LOW	LOW	LOW	LOW
22	MEDIUM	AVERAGE	LOW	LOW	LOW	LOW	LOW	LOW	LOW
23	MEDIUM	SLOW	MEDIUM	LOW	MEDIUM	MEDIUM	LOW	LOW	LOW
24	LOW	AVERAGE	MEDIUM	LOW	LOW	LOW	LOW	LOW	LOW
25	HIGH	FAST	MEDIUM	HIGH	HIGH	HIGH	HIGH	HIGH	HIGH
26	HIGH	FAST	MEDIUM	HIGH	HIGH	HIGH	HIGH	HIGH	HIGH
27	MEDIUM	AVERAGE	HIGH	HIGH	MEDIUM	MEDIUM	HIGH	HIGH	HIGH
28	MEDIUM	AVERAGE	MEDIUM	MEDIUM	MEDIUM	MEDIUM	MEDIUM	MEDIUM	MEDIUM
29	HIGH	SLOW	LOW	LOW	LOW	LOW	LOW	LOW	LOW
30	MEDIUM	AVERAGE	MEDIUM	MEDIUM	MEDIUM	MEDIUM	MEDIUM	MEDIUM	MEDIUM
					17/30	21/30	28/30	29/30	28/30

Fig. 6 Defuzzification method testing

however, as a testing method it still produces accurate enough results to inform and give insights into the necessary steps to improvement.

The expected result is calculated using the ruling system, previously set up within the Fuzzy Inference System. If the defuzzification value is identical to the expected value, the test is a success. Any errors are highlighted in red. As discussed, the testing solution renders inaccuracies, so the validity of this particular testing solution can only be taken as a rough abstraction of the system's performance.

Figure 7 showcases the final results of the sample test data. Based on the large differences between defuzzification methods, using the improper defuzzification

| Case | Gun Damage | Reaction Time | Accuracy | Win Chance | | | | |
				Centroid	Bisector	MOM	LOM	SOM
1	33	0.26	6.3	59.67685	58	50	58	42
2	35	0.29	8.23	63.68536	68	82.26923	100	67
3	28	0.18	7.98	54.19703	53	50	58	42
4	36	0.29	9.12	74.45018	76	82.26923	100	67
5	33	0.23	6.7	64.20919	61	50	58	42
6	35	0.25	4.13	61.16813	69	75	79	71
7	33	0.36	6.01	43.5537	43	37.5	58	17
8	28	0.3	8.61	45.90944	45	37.5	60	15
9	36	0.4	5.57	53.95261	52	50	52	48
10	30	0.2	7.91	61.6432	65	75	84	66
11	33	0.27	8.36	62.0074	67	75	83	67
12	29	0.19	9.45	58.61124	62	82.26923	100	67
13	36	0.21	8.51	80.29515	80	82.26923	100	67
14	36	0.22	10	80.49296	80	82.7	100	69
15	33	0.35	4.85	37.5	38	37.5	58	17
16	33	0.19	6.42	66.95277	69	75	83	67
17	26	0.4	1.51	16.91146	14	3	6	0
18	33	0.22	7.38	66.16051	66	50	59	41
19	26	0.28	2.48	19.95066	20	17.93103	34	0
20	36	0.25	5.64	75.93825	76	75	77	73
21	27	0.24	6.69	42.93372	36	25	31	19
22	30	0.29	1.33	31.64529	29	25	33	17
23	33	0.31	3.97	47.54902	47	37.5	58	17
24	28	0.25	5.19	37.22162	34	25	33	17
25	36	0.16	5.49	75.69201	75	75	77	73
26	36	0.19	6.03	76.83715	76	75	83	67
27	29	0.24	7.71	52.79749	54	75	84	66
28	29	0.24	4.3	41.68054	43	50	58	42
29	36	0.38	1.45	33.39202	30	25	30	20
30	33	0.27	5.81	55.66685	55	50	58	42

Fig.7 Configuration and output values

value is likely to result in impactful flaws, hence it is important to pick the most accurate method. In this instance, the centroid defuzzification method achieves the gradual change this paper aims to demonstrate. Further, slider analysis (as briefly demonstrated in Fig. 8) confirms that centroid allows for the desired nuance. Other defuzzification methods, such as SOM, remove decimal points, as well as snapping to specific ranges which understandably produce rigid and confusing results.

Once the rules were complete, any issues that arose were related to the inherent properties of Fuzzy Logic. For example, the win chance would appear to fall as the accuracy went down, but as the accuracy entered "low", the win chance would increase. This is due to the fact that when the accuracy has a weak correlation with "medium" accuracy the shape produced in the win chance is low, however, when the accuracy has a strong correlation with "low", the shape has a much bigger area. This was fixed by ensuring the overlap between the "low", "medium" and "high" accuracies were overlapping. This provided a gradual change which mapped to the win chance smoothly.

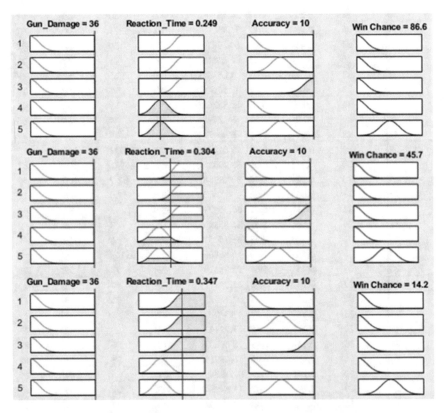

Fig. 8 Three variations of the fuzzy system. The red line can be dragged across the spectrum of each input to alter its value

As demonstrated in Fig. 8, the alteration of the reaction time has a significant impact on the player's Win Chance. It is important to note that the gun damage and accuracy remain the same, as well as the fact that the Win Chance changes dramatically. This result may seem counter intuitive as only the reaction time is changing from "average" to "slow", however, the test data shows that this is the case. When considering the fact that if a player is slightly too slow to start shooting, they will likely die it is easier to understand how impactful a small change can be.

5   The Future of eSports Analytical Tools

If the rising trend of analytical tools in the sports crosses over to eSports, the complexity and accuracy of the systems will rise. It is likely that insights will be gained that humans are currently unaware of, and given that eSports is still regarded by many as in its infancy we will likely see sweeping changes made to the nature of the way in which players approach their training.

Knowing all the variables that impact a player's performance is the ideal state of analytics that would provide the best actionable outcomes, but is knowing every variable an impossibly difficult task?

The debate ultimately lies in the player's willingness to forfeit their privacy. Whilst low hanging fruit such as training hours, ability to perform under pressure, and sleep quality all have a significant impact on their overall performance, smaller factors such as their mood, whether they flew on a plane to get to the competition, and smaller factors still, such as their current relationship with a relative. This list could continue down to the smallest possible variable. Over the next decade, the line between performance and privacy will begin to fade, and it is unlikely the line will be the same for everyone. Analytical tools should aim to allow for the addition of smaller variables whilst ensuring privacy. Without this privacy, it may deprive players of key insights. For example, a player may discover that every time they have a phone call with a parent, they play 3% better the following day due to less emotional stress resulting in higher performance within stressful situations.

In addition to the privacy issue, it may be that a player simply doesn't care enough to add smaller variables and is willing to lose a certain amount of accuracy. This can be remedied with tracking technology integrated seamlessly into the player's life. For example, whenever the player has a phone call, the person who it's with, the call duration and the time of the phone call could be uploaded to the system and automatically integrated for instant insight. The application of future tracking technology will prove useful within analytical tools.

Artificial intelligence and machine learning will impact analytical tools in a series of ways. There are two schools of machine learning that will provide actionable outcomes to eSports athletes. The first is external analysis. This might involve a machine learning algorithm scanning through millions of professionally played eSports games and formulating insights. The insights will range from obvious, such as "better aim provides a better outcome", to almost insignificant, such as "throwing a grenade 0.001 s earlier when playing on a specific map against a specific opponent will provide a better outcome". This external analysis will provide a framework for players to live up to—they will know for a fact the best foundation to play upon. This level of artificial intelligence will be the first stage, and it will equalize the performance of players by pointing out huge flaws and huge performance differences. The second wave of artificial intelligence will be internal. This wave has been talked about in this paper and will involve every aspect of a player's life being fed into the system to provide richer, more personalized insights.

Bio tracking will play an important role in giving personalized insights to assist in player performance. Measuring things like heart rate can already be done with devices that the player attaches to their finger or wrist. As health and medicine are improved by artificial intelligence, the measuring of more intricate health metrics will become relevant to eSports athletes. Their health data can be fed into the artificial intelligence algorithm to provide health insights that will then improve their overall performance.

6 Conclusion

The Mamdani style Counter-Strike Fuzzy System takes in three aspects of a Counter-Strike players gameplay and provides a percentage of the player successfully winning a duel against an opponent. It does so effectively and accurately, with few flaws.

The membership functions used are appropriate for the system. The trimf membership function allows a consistent increase in the outcome and has no significant downsides. This contrasts with the gaussmf membership function which does not provide linear progression. The final solution contains major overlaps between membership functions, which contributes to the success of the system. Without overlap, the Win Chance has the tendency to bob up and down when moving linearly through a single input, which goes against the criteria for the final system.

As discussed earlier, a C++ script was used to rank the rule sequences to apply a subjective ordering of which inputs resulted in which outputs. The fault with many Fuzzy Systems is that subjectivity is considered acceptable because at least some of the system is subjective. A considerable amount of randomness and inconsistency could be removed from systems such as this one by implementing objective standards. The script ordered the rules and assisted greatly in achieving the final result. Without the script there would have been confusion when carrying out tests and the outcome would have been inaccurate.

There are numerous data inputs that haven't been considered in this system due to time constraints and a desire to achieve results with a simple solution. Some examples are shooting style, whether the player is moving, allowing different guns, allowing different body armour, allowing grenades, the strength of a player's position versus their enemy and many more. The vast complexity of Counter-Strike could not be contained in such a simplistic system. Whilst the major aspects have been accurately captured there are many more factors which will meaningfully change the results.

The failure of this system is not in its execution of the data, but in the lack of data it allows.

Designers now have the ability to adapt the framework outlined in this paper. Whilst the system in this paper was created using fuzzy logic within MATLAB, the potential for different technologies is endless. Fuzzy logic has proven to be successful with control systems, however, its applications have yet to be fully utilized in other areas. Fuzzy logic may be an appropriate method of providing a successful chance, however, integrating with other technologies and languages will likely reveal new optimizations and techniques, as is the case with any new field.

References

1. WIN.gg. (2020). *These are all of the biggest Esports prize pools from 2019—DOTA2—News—WIN.Gg* [online]. Retrieved April 13, 2020, from https://win.gg/news/3387/these-are-all-of-the-biggest-esports-prize-pools-from-2019.
2. Business Insider. (2020). *Esports ecosystem report 2020: The key industry players and trends growing the esports market which is on track to surpass $1.5B By 2023* [online]. Retrieved April 12, 2020, from https://www.businessinsider.com/esports-ecosystem-market-report?r=US&IR=T.
3. MarketWatch. (2020). *Sports analytics market 2019 industry news by revenue, business growth, top key players update, industry demand, share, global trend, business statistics and research methodology by forecast to 2024* [online]. Retrieved April 12, 2020, from https://www.marketwatch.com/press-release/sports-analytics-market-2019-industry-news-by-revenue-business-growth-top-key-players-update-industry-demand-share-global-trend-business-statistics-and-research-methodology-by-forecast-to-2024-2019-07-01.
4. https://influencermarketinghub.com/growth-of-esports-stats/. (2020). *The incredible growth of esports [+ Esports Statistics]* [online]. Retrieved April 12, 2020, from https://influencermarketinghub.com/growth-of-esports-stats/.
5. Novatchkov, H., & Baca, A. (2013). Fuzzy logic in sports: A review and an illustrative case study in the field of strength training. *International Journal of Computer Applications, 71*(6), 8–14.
6. Pirovano, M. (2012). *The use of fuzzy logic for artificial intelligence in games.*
7. Singh, H., Singh, G., & Bhatia, N. (2012). Election results prediction system based on fuzzy logic. *International Journal of Computer Applications, 53*(9), 30–37.
8. Voskoglou, M. G. (2014). Assessing the players' performance in the game of bridge: A fuzzy logic approach. *American Journal of Applied Mathematics and Statistics, 2*(3), 115–120
9. Singh, G., Bhatia, N., & Singh, S. (2011). Fuzzy logic based cricket player performance evaluator. In *Artificial intelligence techniques—Novel approaches & practical applications.*
10. Tavana, M., Azizi, F., Azizi, F., & Behzadian, M. (2013). A fuzzy inference system with application to player selection and team formation in multi-player sports. *Sport Management Review, 16*(1), 97–110.
11. Railsback, D., & Caporusso, N. (2018). Investigating the human factors in eSports performance. In *Advances in human factors in wearable technologies and game design* (pp. 325–334).
12. Yang, Y., Qin, T. & Lei, Y. (2016). *Real-time eSports match result prediction.*

13. Joshi, R., Drachen, A., Demediuk, S., Gupta, V., Li, X., Cui, Y., et al. (2019). A team based player versus player recommender systems framework for player improvement. In *Proceedings of the Australasian Computer Science Week Multiconference on—ACSW 2019.*
14. Schubert, M., Drachen, A., & Mahlmann, T. (2016). *Esports analytics through encounter detection.*

Water Carbonation Fuzzy Inference System

William Chapman and Arjab Singh Khuman

Abstract This report will be looking at how fuzzy logic is used to create a system which automatically carbonates water to create sparkling water. This is a topic that has not been discussed a large amount and there is little about it in the associated literature. The system created in this work uses the research available to create a system that carbonates water based on the temperature of the water, the amount of water being carbonated and the sparkling preference of the user. This is then processed in the Water Carbonation Fuzzy Inferencing System (FIS) which outputs to give the Carbon Dioxide Usage. This system is made for domestic water carbonation products and could be extended to larger or smaller products of the type. Several tests have been carried out to measure the success of the system. Changes are then made, and the system is tested again to make sure that the system has been improved. Tests are continued until the system is efficient for the purpose and all the different possible parameters are checked. A critical reflection on the work illuminates the good points, limitations and where improvements could be made.

Keywords Water carbonation · Fuzzy controller · Fuzzy carbonation controller · Carbonation preference

1   Introduction

Here Fuzzy Logic is going to be used to create an automatic water carbonation system. The reason that Fuzzy Logic can be utilised effectively here is that there is a limited amount of research that has been done on this topic allowing Fuzzy Logic to help fill in any issues or lack of literature. What research has been found will be presented in this report, as well as the way Fuzzy Logic has been used to form the system.

W. Chapman
School of Computer Science and Informatics, De Montfort University, Leicester, UK
e-mail: p17161472@alumni365.dmu.ac.uk

A. S. Khuman (✉)
School of Computer Science and Informatics, Institute of Artificial Intelligence (IAI), Leicester, UK
e-mail: arjab.khuman@dmu.ac.uk

Part of this are articles that provide data on the carbonation of water and alcoholic beverages with temperature and pressure influencing how carbonated the liquid can become. There will then be detailed testing covering defuzzification, weighting and operators which will be used to improve the system to be more efficient and reliable. Fuzzy logic is a great approach for this topic as it is not confined by crisp values such as binary logic.

MATLAB, which is a piece of software, was used as it allows the creation of the FIS that is required for the Automatic Water Carbonation to operate correctly. MATLAB also contains a useful FIS toolkit which is used as a quick way to lay out the basis for a FIS. Another piece of software that was used here was Microsoft Excel to create rule matrices, create inputs and record outputs for tests that were ran with the system.

2 Motivation

With the rise in awareness of the amount of plastic that we use and the growing ability for convenience, it is only logical that machines that allow things like the automatic carbonation of water would grow in popularity. For example, one article stated that you can "Make Your Own Sparkling Water And Curb 500 Plastic Bottles Per Year" [7] which could mean that this could help battle climate change if this is adopted in mass. What this review will be looking at is what literature is currently out there, what they discuss and what can be concluded from an analysis of them. The themes that will be covered here will include Fuzzy Logic, Carbonation and overall research found around water dynamics and its carbonation.

The main part of this is using Fuzzy Logic which will, especially with this type of system which lacks the amount of research and literature of other possible systems. This is because Fuzzy Logic can solve any issues of uncertainty if there is not enough data where normal logic would not be enough to work. Something like binary logic can only work with absolute numbers which a solidly defined whereas fuzzy logic allows the use of degrees of membership. This means that anything that can only be approximated can still be effectively implemented and be part of a working system. This is also true for any information that may not be exact, or it might not be consistent as fuzzy sets can deal with these issues. The fuzzy logic techniques described here [15] mean that mapping classes to binary values are also possible if required. There are also various examples of fuzzy logic being used to create systems that are required to make decisions such as [8]. This shows how fuzzy logic can be utilised to create complex systems, in this case, the carbonation system is able to be automated using fuzzy logic. Another thing we can gather from this is how flexible fuzzy logic is and how it can be applied to a variety of systems, especially when we lack certainty or clarification in what the system needs to do.

While being able to use fuzzy logic to create a decision-making automated water carbonation system is incredible, it is also good to have some kind of basis for the system in the form of data or knowledge. In this case, one thing that is useful is a Beer Carbonation Chart [1] which presents the temperature and pounds of pressure per square inch in order to reach to correct volume of carbon dioxide in any given liquid. This provides a good idea of what the inputs need to be adjusted to for this system like that low temperatures mean that carbonation is made easier and high temperatures have the opposite effect. One article looks specifically at the carbonation of water which is exactly what this system is designed to do, and this shows another useful chart [2]. It also has Temperature and pounds per square inch affecting the volume of carbon dioxide in the water. This supports the previous research while also being more reliable as this was done using water instead of alcoholic beverages. A fuzzy system can be built around this kind of data as it gives some structure to it while still allowing fuzzy logic to be used, for example, where the water temperature is perceived to be high, medium or low when being carbonated.

Looking at other supporting information that is relevant to the topic, there was useful research on the effects of temperature on liquids where "the density of a liquid is affected by change in temperature" [6], which explains how higher and lower temperature can change and how much carbon dioxide is needed to properly carbonate water to the amount required. While this article does provide an explanation of how this has been worked out, it does not seem to reference anything to back up these claims which could render the article potentially unreliable. While this can be used to help further understand water dynamics and how that could affect the way this system would work, it is important to be aware of these issues. What was also found was a Portable Automatic Water Carbonator [14] which illustrates that this concept is viable and achievable. The system described in this paper, however, does not include a fuzzy system like the one that will be created here, so while it does show a concept of a Water Carbonator, it is not the same type. For this reason, its relevance may be reduced. Another example is SodaStream's "Fizzi One Touch" (SodaStream [5], The Fizzi One Touch) which automatically carbonates water for people which is closer to the fuzzy inference system that has been made here. While there is not as much detail to the workings of this system it is useful to know that this kind of system is possible. A fuzzy system using Matlab and other pieces of software were used in a review of carbonated drinks to find a similar answer as this report by looking what balance was best including "mouthfeel, flavour and overall liking using biometrics as inputs" [12]. This was done with a variety of different beverages, not just carbonated water but could provide further insight as even "red glasses elicit a perception and expectation of higher carbonation" [12]. This suggests that even the container being used influences people's tastes and preferences. This in conjunction with "audible sound (25–75 Hz) to modify bubble size" [12] being possible. This expands the factors that could be considered and used in testing and in the fuzzy system itself.

Something else that was found that is worthy of note as a warning was research done that found that sparkling water has caused "rare anaphylactic reaction" [3]. The article did state that a patient had suffered from this due to "exposure to sulfites" [3] which were present in the sparkling water that they had consumed. Something as serious as this should be monitored and possibly researched further in order to warn those at risk or find a way to prevent this from happening with any sparkling water this system would create.

Overall what research has been found does have limitations such as possible reliability problems and some with issues of how relevant they are to the project. The main problem is the amount of research that there currently is, and therefore, there needs to be more done to get a broader understanding of this topic. To conclude, there is limited research and literature on this topic so there is only so much that can be gathered and used for this project. However, the literature reported has proved to be useful for gaining a good understanding of the way water carbonation works and how that information can fit into this fuzzy inference system.

3   System Overview

3.1   Design

When looking at Water Carbonation the first thing to do was to narrow down what the system would consider when deciding how much carbon dioxide it would be using. This is in order to get the system to sufficiently carbonate the water to the correct level for users. These parts that the system will use will be the inputs. The first thing that was obvious through the research that had been found was to include the Temperature of the water that is being carbonated as an input. This is because that the research found stated that lower temperatures could hold more carbon dioxide due to the way their molecules react to different temperatures. The lower the temperature, the lower the amount of carbon dioxide that is needed which can be seen from [2]. The next thing that made common sense as an input was the amount of water that was going to be carbonated. The amount of water needs to be measured so that the system can account for differences in potential high or low amounts of water. So, for example, the lower the amount of water that is being carbonated the lower the amount of carbon dioxide that will be used. The last thing that was added to the input was the idea that the user could set a preference for how sparkling they wanted it to be. This gives the user the choice, as users have preferences on how sparkling the water they want. As there is not any crisp way to define the parameters of what the user's preference for sparkling water is, it was added as a percentage from 0 to 100%. This means that 100% is they prefer a higher carbonated drink and 0% is they do not want their drink to be sparkling at all.

For the design of the rule base, the idea would be to have the rule to cover every possible variation of input so that no matter the input there is a rule available to be fired. This is the reason that the operator AND was used so that each rule is very specific. The design for the weighting will be at 1 at the creation of the system as there is no reason for it not the be the default value.

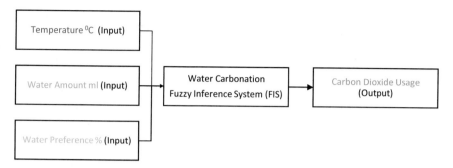

Fig. 1 Diagram of the water carbonation FIS

3.2 Fuzzy Inference System

See Fig. 1.

3.3 Description of System

The Water Carbonation Fuzzy Inference System has four inputs which are used to evaluate the amount of carbon dioxide that the system uses in order to carbonate the water. These inputs are Temperature in Celsius, Water Amount in Millilitres and Water Preference as a Percentage. Celsius was used for Temperature as it is the measurement that is most familiar, Fahrenheit was also considered. Millilitres was used for Water Amount as it is the unit of measurement used with some of the research that was used such as the bottles that are used to carbonate water. The line of the bottle indicates the recommended number of millilitres to use. Percentage was chosen for Water Preference as there is no proper measurement to use for it, so percentage seemed to best fit with fuzzy logic. Lastly, Grams were used for the Carbon Dioxide output as the research that was found for the development of the system uses this measurement, so it makes sense to use it (Table 1).

3.4 Parameter and Interval Justification

The first input temperature was chosen to have a range of between −1 and 20 °C. This was chosen because of the research that was found such as [1] and [2]. This supports the range chosen as anything lower than −1 is not relevant as there is no data for it, probably because the liquid will become a solid meaning that carbonation is impossible. The higher parameter was chosen similarly as the data used for this input stopped at this temperature as anything above is inefficient or not possible for

Table 1 Overview of water carbonation FIS showing input and output variables with their types, ranges and intervals

Variable	Type of variable	Range	Intervals
Temperature (°C)	Input	−1 to 20	Low, medium, high
Amount of water (ml)	Input	600 to 1000	Low, medium, high
Water preference (%)	Input	0 to 100	Low, medium, high
Carbon dioxide usage (g)	Output	0 to 20	Low, medium, high

carbonating water. The second input which is the amount of water has been chosen to have a range of between 600 and 1000 mL. This is the chosen range due to this design being based around current viable commercial products such as SodaStream which currently offers two bottles that are used to carbonate water with their products, containing a maximum of 1000 and 500 mL, respectively (SodaStream [11], Spirit). The third input water preference was chosen with much more emphasis on a fuzzy logic methodology in mind because there is a lack of literature to help with this input. Therefore, this chosen input will have the system to calculate the preference as a percentage and have a range of between 0 and 100%. 0 being no water carbonation preference and 100 being the highest water carbonation preference.

The output of carbon dioxide usage has a chosen range of between 0 and 20 g of carbon dioxide. This again has a bigger use of fuzzy logic due to a lack of crisp data. As the system is partially based on a commercial product through the range of the amount of water, the logical thing to do was to base what average use of carbon dioxide was in those products. The lower end of the range makes common sense at being at 0 where no carbon dioxide is used. However, what was able to be worked out for the higher end from what information there is available was that the canisters used are filled with approximately 60 L worth of carbon dioxide and specifically contains 425 g of carbon dioxide [4]. Using these figures, we can deduce by dividing the grams of carbon dioxide by the litres that it can carbonate that the average amount of grams used per litre of water is when rounded, 7 g. What can also be found is the recommended amount of water by the company, which is 840 mL [13], which is noted as the line on the 1 litre bottle. Using the previous figure of 7 g used for 1000 mL, we can work out that 840 mL should use 5.9 g of carbon dioxide. The company also recommends (SodaStream [10], level of carbonation) how long to have the product carbonating the water and with 8 s being on the high end, we can assume that 4 s is a good average for a time the water is carbonated normally. Finally, taking all these figures into consideration and if 5.9 g is used for the average amount of water over the average amount of time, we get a figure that 1 s of carbonation equates to 1.5 g of carbon dioxide used. Using the higher figure of 8 s and if some may go further as much as 50% more at 12 s, we get the figure of 15 g being the highest probable use of carbon dioxide. The major problem that causes these figures to only be approximate is that all the figures obtained from the company will themselves be an approximation and it also relies on assumption that the average is based on the correct temperature of water between 4 to 8 °C (SodaStream [9], Sparkling Water Maker). Due to these big problems with these calculations, the range is extended to 20 g to consider the possible miscalculations.

The intervals for all inputs and outputs are the same will Low, Medium and High. This is to keep the system simplified as none of the research that was found justified a requirement of having anything other than said intervals. This setup also does make sense given the ranges of all the inputs and any data that will be used for them and it also provides more of a system uniformity.

3.5 Operator, Implication and Aggregation Justification

The operator AND has been used for the entire rule base to make it more focused. The rule base covers every single possible variation of inputs with the AND operator so that there should only be 1 rule that is fully satisfied by all the inputs. With the rule base covering all the possibilities, if the operator was OR, then the number of rules would increase dramatically which would make the system less efficient upon processing. The operator of AND has been set at minimum in the system as this seems to be the values used in a lot of example FIS systems that have been studied in the past. The OR operator has been set at maximum but because there are currently no rules using this operator, it does not make a difference whether it is maximum or minimum. However, it was set at maximum in line with the past example FIS systems that have been studied. The Implication and Aggregation were both picked with each other in mind with the Implication be set at the minimum and the aggregation at the maximum in order to get more average values from the system. This has also been chosen for the same reason as the AND was chosen to be minimum because this being a usual setup for FIS systems. Something to note is that these are parts of the system that could be tested to see whether the different selections would have any significant impact on the system's processing and results.

4 Experimental Design, Testing and Evaluation

4.1 Initial Design

The first design of the fuzzy distribution for the Water Carbonation FIS and the reasons why these membership functions are used were the following. The Trapezoidal (trapmf) was used for all the intervals of Temperature as the data obtained through research gave a solid base to show what was defined as Low, Medium or High. This resulted in plateau regions for all the intervals, which mean, it has a membership association of 1. This allows a more gradual distribution when compared to other types of membership functions. As for the Amount of Water, the leftmost and rightmost intervals both used the Trapezoidal membership function, but the central interval instead used the Triangle (trimf) membership function. This was because it did not need the advantages of the Trapezoidal and can display the overlaps between the intervals better. The Sparkling Preference and Carbon Dioxide Usage are given the Gaussian membership function (gaussmf). The reason the Sparkling Preference has this is because it is more of a fuzzy input in nature being more subjective on what is someone's sparkling preference. For this reason, the Gaussian allows this to be displayed better with a more gradual and flexible change from one interval to the next. With the Carbon Dioxide usage, it has been used to better illustrate how the change in inputs affects the gradient of the output.

4.1.1   General Notes

Some issues that may arise depending on the outcomes of the tests are as follows. The current temperature of the interval goes as low as -1 °C and below that the liquid will become a solid and will no longer be able to be carbonated, this means that if this is the case then the carbon dioxide usage output should give a value of 0. It's a similar situation with the amount of water input as if it is 0 because if there is no water then the system should not use carbon dioxide on anything.

The current number of rules in the system is 27 which is not a large amount, these cover every possible input. This results in 27 rules as there are 3 intervals for each of the 3 inputs. The testing will show whether this is the correct set of rules for the system or anything needs to be removed, change or added.

5   Testing

Dummy data will be used for testing as there are no known real-world tests that have been done that are available to use on this system. A set of 5 test cases were also used before the main testing starts to make sure that MATLAB can successfully read the test data from an Excel file to the .m file. This should then mean that the FIS is successful in outputting the results to another Excel file. This will make sure that not only is the FIS working as currently made, but also that MATLAB and Excel are also working correctly to allow further testing.

5.1   Test 1

Once the 5 test cases were successful in making sure that the FIS, Excel and MATLAB were all working correctly, then the system was ready to use a larger set of test cases. This will consist of a set of 30 randomised values within the ranges of the inputs which

Table 2 Shows expectations of the system

Temperature/Water	Low	Medium	High
Low sparkling preference			
Low	Low	Low	Medium
Medium	Low	Medium	Medium
High	Medium	Medium	High
Medium sparkling preference			
Low	Low	Medium	Medium
Medium	Medium	Medium	High
High	Medium	High	High
High sparkling preference			
Low	Medium	Medium	High
Medium	Medium	High	High
High	High	High	High

have been selected by Excel's RandBetween function which allows the choice of a random number between a minimum integer and a maximum integer. This allows the test data to be completely unbiased for any kind of preference. Thirty sets of values were chosen for the first test with the idea that as the system currently has 27 rules, then the randomised test data has a good chance of showing results of the whole array of the rule base being fired. The dummy data that was generated was used to test the system. Below in Table 2, the systems expected results are displayed.

5.2 System Expectation

Test 1 Data Rules

1. Temperature: Can only be between −1 and 20 °C
2. Amount of Water: Can only be between 600 and 1000 mL
3. Sparkling Preference: Can only be between 0 and 100%.

5.2.1 Test 1 Results

Five defuzzification methods were used to calculate the crisp results of the Water Carbonation FIS. The results of these tests are shown in Table 3 showing out of thirty cases which were correct and incorrect for each defuzzification method.

The results were surprising in the fact that for the majority of the defuzzification methods, more than half of the test cases failed. This means the system is performing poorly to what it is expected to result in. The most successful method was the MOM at 16/30 cases having a result that was correctly expected. It was interesting as it showed the problems such as the temperature and amount of water mentioned in the

Table 3 Test 1 results

Defuzzification method	Correct cases	Incorrect cases
Centroid	15	15
Bisector	12	18
Small of maximum	14	16
Medium of maximum	16	14
Large of maximum	12	18

general notes which could be changed and solved for the next test. Having tested the system and analysed it, the second test can be planned and executed. For the second test, the system has been changed to have multiple new additions including changes to the system expectations. The changes to the system expectations can be seen in Table 4. These include new intervals for the inputs of Temperature and Amount of Water named Freezing and Not Enough Water, respectively, and a new interval for the output of Carbon Dioxide usage. The new interval named No Usage in Carbon Dioxide Usage is to account for the other new intervals. As mentioned in the general notes, this is to fix the problems such as Temperature and Amount of Water reaching low enough that carbonation should not take place at all. Along with these changes, there has also been an addition of 21 new rules to allow the system to deal with the expanded number of intervals and possibilities. This brings the total amount of rules in the rule base to 48. This allows the rule base to continue to cover every possible variation on the inputs and outputs. The range of Temperature, Amount of Water and Carbon Dioxide usage has also been increased to allow for these new intervals and the previous intervals have had their values refined to be better represented.

5.3 Test 2

For this test, the big changes to the system should give a better chance at better outcomes when compared to the expected outcomes. The rule base has been increased because of these additions of new intervals in Temperature, Amount of Water and Carbon Dioxide Usage. Some of the intervals that were already in the system have also been changed to try and improve them. These changes do unfortunately mean that MATLAB will take a bit longer to process the system because of the larger rule base and increased complexity. The expected outcomes of the system have also changed and are shown in Table 4. The rules for the data that is being used for the second test have also been modified because of the range changes in the variables. This means that a whole new set of randomised test cases will be generated.

Table 4 System expectations after test 1

Temperature/Water	Not enough water	Low	Medium	High
Low sparkling preference				
Freezing	No usage	No usage	No usage	No usage
Low	No usage	Low	Low	Medium
Medium	No usage	Low	Medium	Medium
High	No usage	Medium	Medium	High
Medium sparkling preference				
Freezing	No usage	No usage	No Usage	No usage
Low	No usage	Low	Medium	Medium
Medium	No usage	Medium	Medium	High
High	No usage	Medium	High	High
High sparkling preference				
Freezing	No usage	No usage	No usage	No usage
Low	No usage	Medium	Medium	High
Medium	No usage	Medium	High	High
High	No usage	High	High	High

Table 5 Test 2 results

Defuzzification method	Correct cases	Incorrect cases
Centroid	27	3
Bisector	26	4
Small of maximum	26	4
Medium of maximum	29	1
Large of maximum	25	5

5.4 System Expectation

Test 2 Data Rules

1. Temperature: Can only be between −5 and 20 °C
2. Amount of Water: Can only be between 350 and 1000 mL
3. Sparkling Preference: Can only be between 0 and 100%.

5.4.1 Test 2 Results

The five defuzzification methods that were used in the first test were again used to calculate the crisp results of the Water Carbonation FIS. The results of these tests are located in Table 5.

Having changed the system a lot before the second test because of the results showed in the first test, we can see a big change. The test resulted was far more successful in terms of results and the MOM defuzzification methods were the most successful again with 30/30 cases being correct. This means the logical decision here is to make the set the systems defuzzification method to MOM as it had 100% pass rate out of the 30 test cases. This means that the additions such as the new and refined intervals with the extra set of rules in the rule base that were made have managed to fix most of the issues that were problematic in the system.

5.5   Test 3

After having managed to find a suitable defuzzification method for the system to use with a 100% pass rate, the next step is to look at the weightings of the rule base. This is to test whether it is better to have any other kind of weighting other than the default 1 so the best way to test this is to change all the rules in the rule base to a figure of 0.5. This is because there does not seem to be a problem with the current system, so there is no reason to change anything individually. This test is to check whether weighting has any effect on the success of the defuzzification methods. If all the methods improve, then the weighting can be changed, otherwise, the system that was present after the results of test 2 will be the ideal version. The test will again use a new set of randomised data. With the new set of data, the expected results will be different, and the rule base has been altered. The rest of the system remains unchanged from the previous test including the overall expectations of the system. The rules for the data that is being used for this third test are also the same.

Test 3 Data Rules

1.   Temperature: Can only be between −5 and 20 °C
2.   Amount of Water: Can only be between 350 and 1000 mL
3.   Sparkling Preference: Can only be between 0 and 100%.

5.5.1   Test 3 Results

The five defuzzification methods that were used in the first test were again used to calculate the crisp results of the Water Carbonation FIS. The results of these tests are located in Table 6.

The results of test 3 were interesting to analyse has it came out with a few interesting points. The first is that the range of the test data for sparkling preference should have been from 1 to 100%. This is because we can assume that if the user is using the water carbonation system, then their sparkling preference is always going to be at least 1 or above. If the user's preference is 0, then it is illogical that they would be using the system. The other interesting point is that changing the weightings of the rules in the rule base seems to have increased the amount of successful test cases

Table 6 Test 3 results

Defuzzification method	Correct cases	Incorrect cases
Centroid	29	1
Bisector	29	1
Small of maximum	26	4
Medium of maximum	29	1
Large of maximum	25	5

in only a minor way. The minor increase in successes does not seem to be strong enough to support a permanent change in the weighting from before the test at a weighting of 1. The test provides insight but there is no change to the system as a result of the test.

5.6 Test 4

Next, testing the implication and the aggregation makes sense. The system is working correctly with the MOM method, but there is still improvement that can be made with the other defuzzification methods. This test will determine whether the way it was originally set up was correct or whether any improvement can be gained. This will be done by seeing the results of having the implication and aggregation methods changed from what they have been since the initial design. It will see a variation of their implication and aggregation being flipped, both being minimum and both being maximum. Like test 3, if no improvement is found then the system can remain unaltered. The test data rules have had a minor change with the sparkling preference data now having a minimum value of 1 to account for the error that was found in the previous test. Testing the implication at minimum and maximum is not needed, as all the previous tests had this setting, so comparing these results to test 2 makes sense.

Test 4 Data Rules

1. Temperature: Can only be between −5 and 20 °C
2. Amount of Water: Can only be between 350 and 1000 mL
3. Sparkling Preference: Can only be between 1 and 100%.

5.6.1 Test 4 Results

The results of test 4 were very conclusive and are displayed in Table 7. The implication and aggregation of the system were set up correctly at the design stage with the implication at minimum and aggregation at maximum. This test confirmed that this was the correct way to do this system as the results were a failure when compared to the other tests.

Table 7 Test 4 results

Defuzzification method	Correct cases			Incorrect cases		
	Max and Min	Min and Min	Max and Max	Max and Min	Min and Min	Max and Max
Centroid	6	6	6	24	24	24
Bisector	6	6	6	24	24	24
Small of maximum	6	6	6	24	24	24
Medium of Maximum	6	6	6	24	24	24
Large of maximum	6	7	7	24	23	23

Not only was the highest success of a defuzzification only 7 out of 30 cases which were the MOM and LOM methods, but most of the results were the exact same numbers. This means that even the successful cases were not actually calculating the result correctly. The only results to differ from resulting in the same value for each test case was where the Implication and Aggregation were both set at maximum being processed with the Centroid and Bisector method. The results were still incorrect, but the values differed from each other. Overall, this test like test 3, provides some useful information but there have been no alterations to the system because of the results of this test.

5.7 Test 5

The last useful test that can be done on the system to see what results in setting the operator the maximum instead of the minimum that had been set in the design stage of the system. This will change the way every single rule is processed through the Water Carbonation System so a big change is expected. A similar result to test 4 is expected as the system will not be changed unless the system is improved enough.

Test 5 Data Rules

1. Temperature: Can only be between −5 and 20 °C
2. Amount of Water: Can only be between 350 and 1000 mL
3. Sparkling Preference: Can only be between 1 and 100%.

5.7.1 Results

The results were like that of test 4 as this did not improve the results from tests 2 or 3. These results are shown in Table 8.

Table 8 Test 5 results

Defuzzification method	Correct cases	Incorrect cases
Centroid	5	25
Bisector	5	25
Small of maximum	8	22
Medium of maximum	9	21
Large of maximum	10	20

Table 9 Correct case results comparison

Test no.	Centroid			Bisector			SOM			MOM			LOM		
1	15			12			14			16			12		
2	27			26			27			30			24		
3	29			29			26			29			25		
4	6	6	6	6	6	6	6	6	6	6	6	6	6	7	6
5	5			5			8			9			10		

Changing the operator from minimum to maximum reduces the reliability of the results greatly as the highest rate of success was with the LOM defuzzification method at only 10 out of 30 cases being correct. The similarity between tests 4 and 5 is that most of these correct results do not seem to be calculated correctly as there are a lot of identical results despite each case having different input values. This does mean that the test ultimately failed to provide any improvements to be made to the system, so the system remains unchanged from the previous tests.

5.8 Test Comparison

See Table 9.

6 Discussion

After the five tests, there have been multiple changes to the system to fix issues that were found and refine the rule base with the addition of new intervals and changed intervals. The final setup of the system was found to be able to match the expected outcomes very well when using the MOM defuzzification method. Having to add multiple new intervals, change existing ones and add new rules to the system after the first test was difficult to work out for this system. However, after the second and third tests, the system was found to perform well with the changes and MOM was identified as being the best choice for this system. The other defuzzification methods also showed good results after the improvements but the fact that MOM got 100% of test cases correct after the second test was exemplary. The only test case MOM failed in the third test was because of an error in the range that was used in the sparkling preference variable. If this error had not been made, it would have once again achieved 100% success, and therefore, was chosen as the defuzzification method to be used. The fourth and fifth tests were ultimately done to check the weightings, operator, aggregation and implication had been set to the most efficient settings in the initial design. The tests found that settings the system already had after the third test had been done were still the best for this type of system. They provided useful information but did not affect the development of the Water Carbonation FIS as they did not show any improvements compared to the way the system was already set up.

The idea that the rule base, intervals or defuzzification method could be changed after the initial design seemed illogical as the system seemed at a glance to be setup as correctly and efficiently as possible. However, after the first test, it became immediately apparent that there were several things missing that would improve the system and other problems that needed to be tweaked to get the system to more accurately process the inputs to getting the expected results for the output. After the first test, the system rule base increased from 27 to 48 rules. While this resulted in the system taking longer to process, it was done to correctly cover the new intervals that were added as a result of the test findings. A new interval was added to temperature and water amount to take into consideration if not enough water was available to be carbonated or if the temperature was too low that the water became ice, and therefore, not able to be carbonated. The pre-existing intervals were also refined, and the ranges of the variables altered with the research that had been found still supporting these changes. While this ended up allowing the system to work much more accurately which showed in the results in the second test, there were also some downsides. The system became had a larger rule base and the complexity of the overall system had increased meaning that processing was now slower. This had been done to the improve the results of the system in accurately processing the test data. These problems that arose in the first test had been thought of and mentioned beforehand, so the results were not surprising and could be more easily accommodated. The third test was to discover if there were any advantages or disadvantages in changing the weighting of the rule base for the system. In this case, there were small improvements and a minor error in the testing was found regarding the range of the test data for the sparkling preference. This error did not affect the integrity or reliability of the tests or the test data in any meaningful way.

The Carbon Dioxide Usage output was created with Low, Medium and High as its intervals. This was considered fine until considering a situation where no carbon dioxide needed to be used at all. This was confirmed with the results of the first test and so, therefore, the No Usage interval was added. This then meant that new rules would need to be created to use this new interval. The first one was the temperature variable as it originally had Low, Medium and High as its intervals, but this did not consider if the temperature went below $-1\ °C$ was fixed after the first test with the addition of the interval Freezing. Along with this were new rules to indicate that regardless of what the amount of water or sparkling preference was the output should always result in the Carbon Dioxide interval No Usage. This is a similar situation with the Amount of Water variable as it did not account for there being no enough water for the system to carbonate, so the interval Not Enough Water was added. This again meant more rules to say that no matter the temperature or sparkling preference that if the amount of water was below a certain level, then the output had to be no usage.

After all the tests, improvements and choosing to operate on the middle of maximum (MOM) defuzzification method, the system performs very well. It got a 100% success rate in the second test which meant it worked exactly as expected. One of the only possible limitations is that the water amount is limited to the scale used from the SodaStream products which means that it can only carbonate a maximum of 1000 mL. There seemed to be no reason to test whether there was a difference when changing the way, the Water Carbonation FIS processes the minimum and maximum values. This was because the system currently works perfectly as it is after having completed the third test.

7 Conclusion

To conclude, this has been an exciting investigation to apply fuzzy logic to a novel application area. Going through the different iterations of the system was interesting as it was not something that was expected when the first draft of the Water Carbonation FIS was made. It did not seem like anything could be improved but after doing testing there were many changes that needed to be made to make the system much better. The system at the end of this project has been developed into a working FIS that correctly processes data that it is given.

The completion of the work was aided by having had access to multiple examples of FIS in various application areas. A potential future iteration would be to develop this system again with even more resources and time to include additional factors such as the effect and possible use of sound and perception on the consumption of carbonated water.

References

1. Admin, D. (2015, February 21). Beer carbonation chart: The importance of PSI. *Drinktanks* [online]. Retrieved Nov 17, 2019, from https://www.drinktanks.com/beer-carbonation-chart-the-importance-of-psi/.
2. Bonneo. (2015). *The science of carbonation: A visual guide to great carbonation* [online]. Bonneo. Retrieved Nov 17, 2019, from https://www.bonneo.com/blogs/our-creations/190 92875-the-science-of-carbonation-a-visual-guide-to-great-carbonation.

3. Dean, R. K., & Subedi, R. (2018). More than a drink: A rare anaphylactic reaction to sparkling water. *American Journal of Emergency Medicine* [online]. Retrieved March 21, 2019, from https://doi.org/10.1016/j.ajem.2017.10.019.

4. Ebay. (2019). [online]. Retrieved Nov 20, 2019, from https://i.ebayimg.com/images/g/CG8 AAOSwe9VcK0CE/s-l1600.jpg.

5. iNeedTheBestOffer. (2017). SodaStream fizzi sparkling water maker. [Image]. *INeedTheBestOffer* [online]. Retrieved Nov 17, 2019, from https://ineedthebestoffer.com/wp-content/uploads/2017/11/SodaStream-Fizzi-Sparkling-Water-Maker.jpg.

6. Johnson, S. (2017). *Science project: The effects of temperature on liquids* [online]. Sciencing. Retrieved Nov 17, 2019, from https://sciencing.com/science-project-effects-temperature-liq uids-7796706.html.

7. Kaplan, D. (2019). Make your own sparkling water and curb 500 plastic bottles per year. *GreenMatters* [online]. Retrieved Nov 17, 2019, from https://www.greenmatters.com/news/ 2017/08/28/1vDC40/sparkling-water-bottles.

8. Lee, M., Tarokh, M., & Cross, M. (2010). Fuzzy logic decision making for multi-robot security systems. *Artificial Intelligence Review, 34*(2) [online]. Retrieved Nov 17, 2019, from https:// link-springer-com.proxy.library.dmu.ac.uk/article/10.1007%2Fs10462-010-9168-8.

9. SodaStream. (2019). *How can I get the best results from my sparkling water maker?* [online]. SodaStream. Retrieved Nov 20, 2019, from https://support.sodastream.com/hc/en-us/articles/ 360009091594-How-can-I-get-the-best-results-from-my-Sparkling-Water-Maker-.

10. SodaStream. (2019). *How do I get the right level of carbonation?* [online]. SodaStream. Retrieved Nov 20, 2019, from https://support.sodastream.com/hc/en-us/articles/360009219 693-How-do-I-get-the-right-level-of-carbonation-.

11. SodaStream. (2019). *Spirit* [online]. SodaStream. Retrieved Nov 20, 2019, from https://sodast ream.co.uk/products/spirit.

12. Viejo, C. G., Torrico, D. D., Dunshea, F. R., & Fuentes, S. (2019). Bubbles, foarm formation, stability and consumer perception of carbonated drinks: A review of current, new and emerging technologies for rapid assessment and control. *Foods, 8*(12) [online]. Retrieved March 21, 2019, from https://doi.org/10.3390/foods8120596.

13. Walmart. (2019). [online]. Retrieved Nov 20, 2019, from https://i5.walmartimages.com/asr/ 5487edcd-2d8a-49f5-a0e2-9e8e37e917da_1.33549749a48c722604e0ace1713a97af.jpeg?odn Height=450&odnWidth=450&odnBg=ffffff.

14. Warden, R., & Blythe, S. (1993). *Portable automatic water carbonator* [online]. Retrieved Nov 17, 2019, from https://patentimages.storage.googleapis.com/3b/65/c5/41212fedada487/ US5182084.pdf.

15. Yager, R., & Filey, D. (1993). On the reasoning in fuzzy logic control and fuzzy expert systems. In *Second IEEE International Conference on Fuzzy Systems* (pp. 839–844). Retrieved Nov 17, 2019, from https://ieeexplore.ieee.org/stamp/stamp.jsp?tp=&arnumber=327551.

Correction to: An Outlier Detection Informed Aggregation Approach for Group Decision-Making

Chunru Chen, Tianhua Chen, Zhongmin Wang, Yanping Chen, and Hengshan Zhang

Correction to:
Chapter "An Outlier Detection Informed Aggregation
Approach for Group Decision-Making" in:
J. Carter et al. (eds.),
Fuzzy Logic, **https://doi.org/10.1007/978-3-030-66474-9_7**

The book was inadvertently published with chapter author's incorrect name. This information has been updated from "Tianghua Chen" to "Tianhua Chen" in the initially published version of chapter "7". The chapter and book have been updated with the changes.

The updated version of this chapter can be found at
https://doi.org/10.1007/978-3-030-66474-9_7

© Springer Nature Switzerland AG 2021
J. Carter et al. (eds.), *Fuzzy Logic*,
https://doi.org/10.1007/978-3-030-66474-9_16

Printed in the United States
by Baker & Taylor Publisher Services